FOOD AND DRUG LEGISLATION
IN THE NEW DEAL

FOOD AND DRUG LEGISLATION IN THE NEW DEAL

CHARLES O. JACKSON

PRINCETON UNIVERSITY PRESS

PRINCETON, NEW JERSEY

1970

Publication of this book
has been supported by grants
from Georgia College at Milledgeville
and from the Whitney Darrow Fund
of Princeton University Press

This book has been composed in Linotype Caledonia
Printed in the United States of America
by Princeton University Press, Princeton, New Jersey

To Holly, Tracy, and to the green leaves
who will understand

PREFACE

IN June 1938 Franklin D. Roosevelt signed into law a new Food, Drug, and Cosmetic Act. This law and the Fair Labor Standards Act of the same year represent the last major domestic measures passed under the auspices of the New Deal before foreign affairs absorbed the chief energy of the Administration. The 1938 statute was the first general revision of food and drug legislation since the passage of the 1906 Wiley law. The new act altered considerably the relationship between the federal government and the affected industries. It also eliminated serious and long-standing areas of abuse in the food and drug market. Yet, for all the consequences involved, the legislation has attracted little scholarly attention, even in monographic studies of the New Deal.

The present volume seeks to fill that void through an analysis of the long and complex struggle for what became the Food, Drug, and Cosmetic Act of 1938. On one level the work is a political case study of the legislative process on a highly controversial bill during the New Deal. At its main level it describes a major segment in the social history of federal regulation of the food and drug market. The story itself is a significant episode in the greater overall history of government's reaction to the rapid expansion of science in our day. In this sense the struggle for the 1938 act is more than a matter of one law. It is a statement on the significant and dramatic impact of science upon twentieth-century thought as well as institutions. It reflects concretely the way in which modern scientific, industrial, and political forces interact to affect the course of American social life. I can only hope that I have been able to do justice to the import of the event described.

The major research for this volume was done during a period when I held a U.S. Public Health Service fellowship (1-F1-GM-23, 166-01A1) from the Division of General

Medical Sciences. The book is a direct result of that generous aid. Additional support was also provided by the Faculty Research Committee at Georgia College. I am indebted to several repositories whose staffs extended me every courtesy. Some of these debts are indicated in my bibliographical essay. Special thanks should go to the staff of the Georgia College Library for their continuing patience and efforts to gain materials for me from a variety of locations. My major sources of information were the Food and Drug Administration's correspondence and records for the years 1933-1938. Without the cooperation of the FDA and its permission to consult archival records not barred to me by law the present volume would not have been possible.

A number of individuals assisted me at various stages in the development of the manuscript. My colleague and friend, Professor James Bonner at Georgia College certainly deserves mention for his continuing encouragement to me in bringing the manuscript to completion. I have benefited from the wisdom of his advice and from the scholarly example he set. My greatest debt, one far too large to repay, is to Professor James Harvey Young of Emory University. It was he who first introduced me as a graduate student to the broad subject of this book. He has given me many invaluable suggestions on the preparation of the manuscript since that time. As chief critic and warm friend he is largely responsible for such strengths as this volume possesses. I hope that its publication will serve as a small tribute to him.

I wish also to thank Miss R. Miriam Brokaw and Mrs. Polly Hanford of Princeton University Press for their interest and assistance in my project.

My deepest obligation, in every possible way, is of course to my wife Emma.

CHARLES O. JACKSON

University of Tennessee
Knoxville
December 6, 1969

ACKNOWLEDGMENTS

QUOTATIONS from the Franklin D. Roosevelt Papers, the Josiah W. Bailey Papers, and the Royal S. Copeland Papers are used by permission of the repositories respectively indicated: Franklin D. Roosevelt Library, Hyde Park, New York; Duke University Library, Durham, North Carolina; Michigan Historical Collections, University of Michigan, Ann Arbor. Permission to quote from the Swann Harding Papers was provided by the Manuscript Division of the Library of Congress, Washington, D.C. and by Mr. Swann Harding.

The National Library of Medicine, Bethesda, Maryland has allowed me to quote from "Tugwell, Rexford G., An Interview with Charles O. Jackson, June 7, 1968" located in the Oral History Collection of that repository. The *Journal of the History of Medicine and Allied Sciences* has permitted me to use in revised form material I published previously under the title, "The Ergot Controversy: Prologue to the 1938 Food, Drug and Cosmetic Act," 28 (July, 1968), 248-257.

CONTENTS

FOOD AND DRUG LEGISLATION
IN THE NEW DEAL

I

THROUGH THE LOOKING GLASS

Probably the proximity of the Sylvanian Theater to the Department of Agriculture where they played Alice in Wonderland brought the thought for the preparation and introduction of this bill.

—H. B. THOMPSON, COUNSEL FOR THE
PROPRIETARY ASSOCIATION,
DECEMBER 1933

WALTER CAMPBELL, Chief of the Food and Drug Administration, was irritated. It was a winter day in 1933, and he was on his way across the street to see the new Assistant Secretary of Agriculture, Rexford Tugwell. Tugwell had returned to the FDA office a routine letter, prepared there for his signature, on tolerance levels of fruit spray residue. Attached was a curt penciled comment to the effect that if lead arsenate was a poison why didn't FDA prohibit its use altogether. Campbell's assistant Paul Dunbar was in the office when his chief read Tugwell's notation. The terse comment, Dunbar later recalled, was like "a kick in the teeth" after so many long years of fighting a lone battle against spray residues. So now Campbell was irritated, and he was on his way to tell the Assistant Secretary "a thing or two."[1]

Certainly Campbell would prohibit lead arsenate residue, if he could.[2] No one seemed to understand. There were many things he would do, if only the law allowed, but it did not. His bureau operated under the provisions of a legal statute now twenty-seven years old with very little

[1] Paul B. Dunbar, "Memories of Early Days of Federal Food and Drug Law Enforcement," *Food, Drug, Cosmetic Law Journal* 14 (February 1959), 134. *Hereafter cited FDC Law Jnl.*
[2] *Ibid.*

[3]

having been done to update it. Granted that the 1906 Pure Food and Drugs Act *had been* a good law. Perhaps that was half the problem. Too many people thought it was too good. Even Harvey Wiley who had fought the battle for the first statute came to insist that it was little short of perfect. Wiley, as principal enforcement officer for five years, had waged constant warfare with Presidents Roosevelt and Taft. He had ultimately resigned in 1912, having never converted his superiors to his own militant views on administration of the law. But Wiley argued that the problems in law enforcement were administrative tangles. As late as 1928 he could still say that "there is absolutely no need of any further legislation in addition to what is now on the statute books."[3]

Walter Campbell certainly knew better. There were now serious shortcomings in the old statute. For one thing it did not cover cosmetics, which was a booming business, legitimate and otherwise. A bill before Congress in 1897 had included cosmetics within the definition of drugs, but this language had been dropped in 1900 as partial payment for support of industry groups in a National Pure Food and Drug Congress, organized to develop pressure for enacting a law. It did not seem much of a loss at the time; the great growth of the cosmetic industry was to come after 1906.[4]

Nor did the old law provide adequate control over patent medicines. The Wiley statute was far outdated in its definition of dangerous drugs, for many new products, such as barbiturates, had since been developed or come into wider use. In 1912 Congress sought to correct an adverse court decision on the statute by passing the Sherley Amendment, the language of which proved especially unfortunate. Misbranded items were defined as those bearing statements regarding curative or therapeutic effects which were "false and fraudulent." Thereafter the government was saddled

[3] Oscar E. Anderson, "The 1906 Pure Food and Drugs Act," *The Government and the Consumer: Evolution of Food and Drugs Laws* (Washington, 1963), 6.
[4] *Ibid.*, 8.

with the exceedingly difficult task of proving fraudulent intent in every misbranding case.

The old statute was vague and ambiguous in its language regarding adulteration of food. What the government really needed, and did not have, was authority to set for food products quality and identity standards that would carry the force of law. The 1897 bill had set up machinery for establishing such standards. The final law omitted these provisions, however, and even previously held powers to "suggest" standards were lost in 1907 when the standards clause was dropped from a fiscal act for that year. It was true that the McNary-Mapes Amendment of 1930 allowed the Secretary of Agriculture to set *minimum* standards of quality and fill for canned goods, but minimum standards were inadequate.[5]

Perhaps the single most significant weakness of the 1906 law was that it did not provide for control over false advertising. Not only had Congress failed to invest this power in the FDA, but by 1933 the consumer was virtually without any protection in this area. The Federal Trade Commission, originally charged with responsibility over false and misleading advertising, was itself greatly hamstrung after 1931. In the Raladam Case of that year the Supreme Court ruled that FTC could act only where false advertising was clearly injurious to competitors. It was not enough that the consumer be deceived. For the food and drug market this meant that false therapeutic claims, open to legal action under the Sherley Amendment when printed on labels, could be made with considerable safety in all forms of nonlabel advertising.[6]

The 1906 law was inadequate, and flagrant abuses in the market were growing. A large number of the proprietary panaceas bought by the American consumer each day were

[5] *Ibid.*

[6] For information on the problems faced by FTC see "The Federal Trade Commission Act of 1938," *Columbia Law Review* 39 (February 1939), 259-73; Milton Handler, "The Control of False Advertising under the Wheeler-Lea Act," *Law and Contemporary Problems* 6 (Winter 1939), 91-110.

useless and in certain conditions dangerous. One such nostrum widely advertised and used was Crazy Crystals, represented as a cure for stomach troubles, colitis, diabetes, and other diseases. "Crazy Crystals Pulled Me Out of the Grave" heralded a testimonial circular. In truth the nostrum was essentially nothing more than Glauber's salt, a cathartic which could be purchased at any drugstore for forty cents a pound. Continued use of the crystals, even by the healthy, could be harmful. "Crazy Crystals are wonderfully effective," an Iowa physician pointed out, "in rupturing the appendix. I had one where the solution ran out of the abdomen after drainage and recrystalized on the sheets."[7]

FDA once seized consignments of the crystals under the Sherley Amendment, but the company simply shifted its claims from the labels to other advertising forms and profitably continued to defraud the public.[8] The advertising-control void was of course an open invitation to abuse. Even the venerable old proprietary Lydia E. Pinkham's Vegetable Compound was widely sold under a label which read simply, "Recommended as a vegetable tonic in conditions for which this preparation is adapted." Its advertising matter, however, asserted the balm's value for sundry female disorders and nervous troubles.[9]

The obesity cures also provided a fertile field for abuse. One was Marmola, which contained a thyroid preparation. Introduced into the system indiscriminately, thyroid was apt to produce radical changes in the condition of the human body and thyroid intoxication. The Federal Trade Commission sought to take legal action against the product on the basis of its advertising claims. The result, however, was the earlier mentioned Raladam decision in 1931 which

[7] FDA publication, *Outstanding Provisions of the New Federal Food and Drug Act*, no. 3 (1933), located in FDA Records, Correspondence on Legislation, Box 3121, Decimal File .062, Records Group 88, National Archives, Washington, D.C. This material is hereafter cited as Correspondence, Box ———.

[8] *Ibid.*

[9] *Ibid.*

virtually emasculated the consumer protection efforts of FTC.[10]

More vicious than Crazy Crystals or Marmola were such cure-alls as Radithor and B & M External Remedy. Radithor was a radium water advertised for the treatment of 160 diseases. Its ingredients could be highly dangerous, even fatal. In at least one documented case, that of a wealthy Pittsburgh businessman, its radioactive components had caused death by bringing about the disintegration of bones in the head.[11] B & M was originally a horse liniment, composed mainly of ammonia, turpentine, water, and egg. It came to be offered to the public as a treatment for human tuberculosis, pneumonia, and other diseases. The Food and Drug Administration collected records of sixty-four fatalities laid to the direct or indirect effects of B & M, as part of a legal case against its producers. Even at this, it took FDA ten years and a cost of $75,000 to prosecute the case successfully.[12]

The abuses within the cosmetic and food trades might be less spectacular, as a rule, but they were just as real. Koremlu, the "safe" depilatory cream, included as one active ingredient thallium acetate, a well-known rat poison. Koremlu, Inc. boasted that its advertising was carried in the "best" magazines. This was basically true. It had also made the pages of the *Journal of the American Medical Association*, which reported nearly twenty cases of thallium poisoning in two years from the use of the depilatory. In July 1932, the Koremlu company, with $2,500,000 in damage suits against it, went into bankruptcy. The depilatory was gone, but no thanks to the Food and Drug Administration. Cosmetics were beyond the legal scope of that agency.[13]

[10] *Recommendations from the Consumers' Advisory Board of the NRA for Code Revision*, "Legislation and Enforcement," located in Records of the Office of the Secretary of Agriculture (Under Secretary Tugwell), Accession No. 1074, RG 16, NA 3.

[11] *Ibid.*, 2.

[12] *Ibid.*, 4.

[13] Arthur Kallet and F. J. Schlink, *100,000,000 Guinea Pigs* (New York, 1933), 80, 84, 87.

Then there were the food preservatives. As the founders of Consumers' Research put it, with perhaps a bit too much melodrama, "the hamburger habit is just about as safe as walking in an orchard while the arsenic spray is being applied."[14] The meat preservative sodium sulphite was a case in point. The sulphite was a hazard to the kidneys and the digestive process. FDA understood these problems. Quantities of chopped meat preserved with sodium sulphite, which restored a fresh red color, were constantly being seized by food inspectors, federal and state. But so easy and profitable was the fraud and so slight was the punishment for violation that the practice remained common.[15]

These facts were general knowledge to Walter Campbell. Perhaps that was the real reason for his irritation as he walked toward Secretary Tugwell's office. He had long wanted to strengthen the law but had gained little Congressional support. The only significant amendments to the 1906 statute since its passage were the earlier mentioned Sherley Amendment of 1912 and the McNary-Mapes Amendment of 1930, along with the Net Weight Act of 1913. The latter required a declaration of quantity on packaged foods.[16] What had been more typical, during the 1920s at least, was a series of so-called ripper bills aimed at reducing FDA's authority, especially its power to make an unlimited number of seizures in the marketplace of products adjudged in violation of the 1906 law. FDA was always on the defensive. Dr. Wiley, with his respected reputation, did not help much either. He had been very critical of the operation of FDA during the decade before his death in 1930, and this attitude was apparent in his writings.[17]

[14] *Ibid.*, 38.   [15] *Ibid.*, 34.
[16] Ruth Lamb, *American Chamber of Horrors* (New York, 1936), 11; Miss Lamb lists a fourth amendment as well. This was the "Shrimp Amendment" of 1934 which authorized supervisory inspection of the seafood industry for all packers desiring the service. This amendment has been omitted from the body of the text since it was passed after the introduction of Senate bill 1944, the first effort at total revision of the 1906 statute.
[17] Harvey W. Wiley, *An Autobiography* (Indianapolis, 1930); *History of a Crime against the Food Law* (Washington, 1929); and

The tide of popular opinion was apathetic if not hostile to FDA's regulatory activities in the twenties. That was bad enough. Even worse, it had proved far too easy for some opportunist to generate criticism and doubt regarding the operations of the agency. Certainly Campbell would never forget the name Howard K. Ambruster. If by chance the name did slip his mind the drug chief need only glance over the constantly growing FDA correspondence file on Ambruster. The whole painful memory would quickly return. As Campbell later wrote, "he [Ambruster] has regarded it as his special mission in life to attack the Food and Drug Administration."[18]

The beginning of that story went back to the fall of 1927. Ambruster was known in the chemical industry as an expert on pesticides and insect control. In 1927 he also became a dealer in crude ergot, a drug used in extract form to combat postpartum hemorrhage. He had telephoned the editors of *Drug Markets* to make this fact known, as well as his asking price, which was slightly more than twice the current market price. The editors were surprised. How, they queried, did he expect to get so much for his crude material? Was the available market supply low? "You had better check," he replied, "I don't think there is any around." So they did. The result of their inquiry was published in the September issue of *Drug Markets* under the title, "A Corner in Ergot."[19]

Ambruster failed to make the quick profit he expected from his presumed monopoly of crude ergot. This fact was the crux of the problems which FDA faced for the next three years. Ambruster had misjudged the current market need as well as the ability of the large drug manufacturers to hold off buying his high-priced Spanish material until the arrival of cheaper Russian ergot around the first of the year. He insisted Russian ergot was inferior to that imported from Spain, but whether this was true or not, he

"A Criticism of Some Drug Law Regulations," *Am. Druggist* 80 (October 1929), 60ff.

[18] Walter Campbell to Paul Appleby, Correspondence, Box 16.

[19] Cited in "Ergot to Ether to Digitalis," *Drug Markets* 26 (June 1930), 569.

could clearly see his profit collapsing with the new imports. He knew also that he could not long hold his own material in its existing form, for crude ergot had a rapid deterioration rate. He might, of course, drop his own price, but that would be to admit that he had sought originally to make an excessive profit.

The alternative was to turn his crude material into extract form, which was more durable, and then to compete directly with the manufacturers who had rejected his goods. If he could also persuade the public that his extract was superior to other extracts on the market, he might even force those manufacturers to buy from him or else risk losing their own retail outlets. The logical means to this end was to reinforce his original claim that Russian ergot was inferior by asserting that substandard, even impotent cargoes, mainly of Russian origin, were actually being processed into extract for the American market. One could only be *sure*, then, by buying Ambruster extract.[20]

The obvious corollary of all this was that Walter Campbell's Food and Drug Administration had been woefully negligent. It was that agency which presumably had allowed adulterated material to enter the country in disregard of the 1906 law. Ambruster launched these charges with zeal about the beginning of 1928.[21] Working through the Confidential Trade Information Bureau, run by Manuel DeCastro, he sought to insert his allegations against FDA in various state medical journals as paid advertisements. This attempt failed primarily because the publishing bureau of the American Medical Association advised against their acceptance.[22] Through personal speeches, and an

[20] *Ibid.*, 570; *Administration of the Food and Drugs Act, Hearings before the Committee on Agriculture and Forestry*, U.S. Senate, 71st Cong., 2nd Sess. (Washington, 1930), testimony by John Vaughn, 678. Hereafter cited as *Senate Hearings on Food and Drugs Act* (1930).

[21] "A Flier in Ergot," *JAMA* 90 (January 14, 1928), 121; *Senate Hearings on Food and Drugs Act* (1930), 226, 576, 925.

[22] *Ibid.*; "Ambruster, Rusby, and Ergot," *JAMA* 95 (September 6, 1930), 722. Ambruster denied any financial relationship with De-Castro.

elaborate distribution of circulars, however, Ambruster did get the charges before a wide audience. By April 1929, his circular campaign against FDA had progressed to the degree that he was propagandizing such diverse groups as the National Better Business Bureau, the Daughters of the American Revolution, and the Federal Council of the Churches of Christ in America. The scope of the charges was also widened by his assertion that FDA had also permitted faulty ether, digitalis, and other drugs on the marketplace.[23]

The plausibility of the charges was materially enhanced by the vigorous support of Dr. Henry Hurd Rusby, Dean of Columbia University College of Pharmacy and a prominent figure in national drug circles. Why such a renowned pharmacologist and a one-time consultant to Harvey Wiley's old Bureau of Chemistry should have become involved in the ergot intrigues is difficult to say. He denied any financial arrangement with Ambruster and attempted, though unsuccessfully, to sue *Time* magazine when it advanced such a thesis. It was true, however, that in recent years Rusby had lent his name to several surprising causes, including the famous patent medicine Wine of Cardui. It is also worth noting perhaps that Rusby's association with the Bureau of Chemistry had not always been a happy one. Charges had once been made that Rusby was illegally overpaid for services rendered to the Bureau. The doctor was exonerated by a special investigating committee but the committee did recommend Rusby's dismissal.[24] The AMA made it clear that they had lost all confidence in Rusby's medical opinions.[25] Whatever the reason, Rusby's name

[23] *Ibid.*, 724; *Senate Hearings on Food and Drugs Act* (1930), 1612, 1571, 1286-88, 580.

[24] *Ibid.*, testimony by Rusby, 129; also see testimony by Dr. Olin West, 297, and Campbell, 611; "Ergot Controversy," *Time* 13 (April 15, 1929), 42; news item, *JAMA* 93 (July 27, 1929), 288. Significantly, Rusby was also for a time on the editorial staff of DeCastro's News Bureau publication. On Rusby's difficulties with the Bureau see Oscar E. Anderson, *The Health of a Nation* (Chicago, 1958), 244-45.

[25] *Senate Hearings on Food and Drugs Act* (1930), testimony by West, 297.

provided unwarranted dignity for the attack on FDA and provided it an entree into circles which might otherwise have been closed. Thus Rusby was given the occasion to introduce the Ambruster thesis into the 1929 convention of the American Pharmaceutical Association, although Ambruster was refused permission to speak to that body.[26]

The surprising thing was that the charges did not enlist more support than they did. With very few exceptions, they received little backing among organized medical and pharmacological bodies. The AMA *Journal* was openly hostile to both Ambruster and Rusby. Reaction in the lay press was brief, sporadic, and of the news variety. Yet, if the extent of concern with the allegations was limited, the support obtained was not without influence. It was certainly sufficient to cast a cloud over the integrity of the Food and Drug Administration at a time, as Walter Campbell pointed out, when popular support for "our work" was never more requisite.[27]

The support for the Ambruster case included no less a body than an official investigation committee of the American Association of Obstetricians, Gynecologists, and Abdominal Surgeons headed by Dr. Edward Ill of New Jersey.[28] Equally important, the cause was taken up by Dr. Harvey Wiley, ex-chief of the Bureau of Chemistry and architect of the 1906 food and drug law.[29] These sources could hardly be dismissed as commercially interested and were undoubtedly highly influential in gaining the support of others within drug circles. Both Ambruster and Rusby insisted by 1930 that they had received many letters of concern from physicians, chemists, and pharmacologists over the country. While the number was likely exaggerated

[26] *Ibid.*, 152; *Oil, Paint and Drug Reporter* 117 (June 16, 1930), 21. *Hereafter cited OP&D Reporter.*

[27] Campbell to Charles McNary, May 13, 1930, in Records of the Office of the Commissioners, Box 16, FDA Records Group 88, National Archives. Hereafter cited as Commissioners' File, Box ——.

[28] The report of this body is summarized in *Congressional Record*, 71st Cong., 2nd Sess. (May 1930), 9266-67.

[29] For specific charges, see Wiley, "Criticism of Some Drug Law Regulations," 60ff.

there is every reason to believe that the expressions of concern were quite real.

The case was also taken up by the newly organized Consumers' Research organization which frequently criticized the FDA in its *General Bulletin*. The organization's founder, F. J. Schlink, was still repeating the ergot charges by 1933 in his sensational muckraking volume *100,000,000 Guinea Pigs*, which so stirred the public in the thirties.[30] The allegations received further support from the magazine *Plain Talk* in the spring of 1930. The June issue of that journal carried an article by Senator Wheeler of Montana filled with reckless and shocking assertions about the current condition of the drug market.[31]

Wheeler's participation was the last straw for Walter Campbell. To allow that type of attack from Congressional circles would never do. He immediately requested a full-scale investigation of his agency by the Senate Agriculture Committee.[32] It was time for FDA to defend itself. No, Walter Campbell was not likely to forget Howard K. Ambruster, nor, for that matter, the "two or three day" investigation that in fact droned on for a month. The ironic thing was that, while Campbell probably could not recognize it even in 1933, the ergot hearings helped to bring about a genuine change of fortune for his agency. There were already some signs of this change on the eve of the hearings. In May the District of Columbia Court of Appeals rejected a judicial assault by Ambruster and upheld the discretionary powers of FDA over what drug materials entered the American market.[33] One trade journal promptly noted its significance: i.e., that those unhappy with the decisions of the agency had better spend their time in working for revi-

[30] James Corbett, "The Activities of Consumers' Organizations," *Law and Contemporary Problems* 1 (December 1933), 63; Kallet and Schlink, 142-56.
[31] Burton K. Wheeler, "Profiteers in Poison," *Plain Talk* 6 (June 1930), 675-83.
[32] Campbell to McNary, May 13, 1930, Commissioners' File, Box 16.
[33] Ambruster v. Mellon, in *Federal Reporter*, 2nd Series, Vol. 41, Court of Appeals, Washington, D.C. (1930), 430-32.

sion of the law since to contest decisions in the courts would be futile.[34]

Then came the June hearings. They ran on for week after week, often without direction and at times bordering on chaos. Ergot was frequently lost in a maze of personal gripes and grievances over everything from Jamaica ginger to creosote. But, at the end, and even though no formal report was written, there could be little doubt about FDA's vindication. The charges and even the testimony of Dr. Ill, as one contemporary put it, were "riddled by the incontrovertible documented facts presented by officials of the Food and Drug Administration."[35] Senator Royal S. Copeland of New York, a participant in the probe, wrote Campbell an open letter, given wide circulation, in which he flatly declared that FDA had come out "with flying colors."[36] Various interested publications such as the *Oil, Paint and Drug Reporter* and the *New York Journal of Commerce* also told their readers there was no longer an issue.[37] The hearings thus cleared the air not merely for the present but for the future.

More important, while FDA was exonerated of failings, the 1906 law was not. Much too often Campbell and his men were able to meet accusations at the hearings by simply pointing to the law's inadequacies. Senator Wheeler, a member of the committee, demanded to know why *all* poor grade ergot and ether should not be cleared from the market. Campbell pointed out that all a manufacturer had to do to get beyond the powers of FDA was to specify on his labels a variation from United States Pharmacopeia standards.[38] To assertions that FDA had returned confiscated drugs to their owners there was the same answer: this was provided for under the law.[39] Wheeler objected

[34] "The Ergot Decision," *Drug Markets* 26 (May 1930), 438.

[35] T. Swann Harding, *Fads, Frauds, and Physicians* (New York, 1930), 333.

[36] Copeland to Campbell, September 3, 1930, Commissioners' File, Box 16.

[37] *Senate Hearings on Food and Drugs Act* (1930), 1892-95.

[38] *Ibid.*, testimony by Campbell, 710.

[39] *Ibid.*, 407.

to the food and drug unit's informal "persuasion" efforts with manufacturers. Strict enforcement and prosecution should be the rule. Campbell again fell back on the law. Its minimal penalties—$200 for first offenses—were hardly deterrents. At this point even Wheeler admitted a need for change in the law.[40]

So the evidence piled up—evidence that if something was wrong in FDA it was the weak law under which the agency operated. The ergot controversy spurred some recognition, even within the drug industry, of the need for revision of the law. The two sequential resolutions passed by the June 1930 convention of the American Pharmaceutical Manufacturers Association were more than a coincidence. The first declared the Association's confidence in the operation of the Food and Drug Administration; the second, its support for federal legislation that would give the Department of Agriculture control over the advertising of drug products.[41] Perhaps some of the delegates had not forgotten Ambruster's advertising campaign for his extract.

Also of major future significance was the fact that the ergot investigation helped to spur the interest of Senator Royal Copeland in the activities of FDA. Because he was a physician and a one-time health commissioner of New York City, Copeland had been invited to sit with the Senate Committee during the hearings. By the end of that probe Copeland's admiration for FDA was so apparent that he was publicly dubbed in the press as "counsel for the defense."[42] The Senator was no one's picture of an ardent reformer. He had come to New York from Ann Arbor, where he had once served as mayor. He held an M.D. degree from the University of Michigan Medical School and had practiced Homeopathic medicine in both Michigan and New York. With the backing of New York City's Tammany Hall he was elected to the United States Senate in 1923. Copeland was not noted for his liberalism, particu-

[40] *Ibid.*, 409.
[41] *OP&D Reporter* 117 (June 16, 1930), 21.
[42] *Senate Hearings on Food and Drugs Act* (1930), 1892-93.

larly in economic matters. He was a jaunty, debonair man, perhaps best known in the Senate for the fresh red carnation he wore in his lapel each morning. In debate he generally gave the impression of being much more anxious to compromise than to stand firm.[43] But the dapper Senator did believe in food and drug reform and in Walter Campbell's agency.

Moreover, Walter Campbell apparently came to believe in the New Yorker. If the Senator was not a flamboyant proponent of the interests of FDA, at least Campbell could be sure he was an earnest one. Later, in 1933, after Campbell had made his visit to Tugwell and a decision was made to seek revision of the old food and drug law, the association with Copeland would be highly significant. Food and drug legislation was a political hot potato. No one wanted to sponsor a new bill, including the respective chairmen of the Senate and House Agriculture Committees, the normal sponsors for such legislation. Copeland came forward and volunteered to take the bill.[44] It was to occupy much of his time in the last six years of his life.

When Walter Campbell made his trip to Tugwell's office in the winter of 1933, it was not very likely that he was in a philosophic mood, and the forthcoming long association with Copeland was in the future. At that moment he was concerned about the Assistant Secretary's curt comment regarding spray residue. The ergot matter probably never entered his mind. If Campbell had been reflective, however, he might have sensed that some good had come out of the public airing of the charges on ergot. If he had been reflective he might have mused that the tide of consumer lethargy seemed to be turning a bit. It was not a striking change, but there were signs.

The first real sign was the publication by Stuart Chase

[43] James Harvey Young, *The Medical Messiahs* (Princeton, 1967), 164; campaign pamphlet, *Royal Copeland for Mayor*, Papers of Royal S. Copeland, University of Michigan Historical Collections, Ann Arbor.
[44] Ruth Lamb to James Rorty, April 11, 1934, Correspondence, Box 438; Lamb, *American Chamber of Horrors*, 285-88.

[16]

and F. J. Schlink of *Your Money's Worth* in 1927. Its theme was the waste of the consumer's dollar which resulted from his ignorance while purchasing goods in the "jungle" of competitive advertising and sales pressures.[45] At the time the book did not seem a landmark. As it began to go into extra editions, however, some advertising men began to wonder. C. B. Larrabee of *Printers' Ink* later termed the volume "some rather brutal and significant handwriting on a rather large wall."[46] The typical reply from the business community was "T'ain't so" with the corollary cry that the work was Bolshevik inspired, if not subsidized by the Third Internationale.[47] Meanwhile, *Your Money's Worth* became a best seller and by the mid-1930s had run through nineteen printings.[48]

The Chase-Schlink work in turn became the prototype for the body of so-called guinea pig muckraking books of the 1930s. Its immediate successor was *100,000,000 Guinea Pigs* by Schlink and Arthur Kallet in 1933. This was a shocking exposé, highly sensational, perhaps overly so, of the food, drug, cosmetic, and advertising market. "In the eyes of the law," the authors charged, "we are all guinea pigs, and any scoundrel who takes it into his head to enter the drug or food business can experiment on us." The legal forms of consumer protection had simply failed, and the so-called integrity of profit-minded manufacturers was worthless advertising chatter. The book concluded with the admonition to "let your voice be heard loudly and often."[49] The number who read that injunction was by no means small. By 1936 the volume was in its thirty-first edition.[50] The list of such publications swelled from 1930 on.

The crusade for the 1906 Wiley statute was materially aided by the journalistic muckrakers of that era and the

[45] Corbett, "Activities of Consumers' Organizations," 61.

[46] Larrabee, "Guinea Pig Books," *Printers' Ink* 175 (April 16, 1936), 72. Hereafter cited *PI*.

[47] Corbett, "Activities of Consumers' Organizations," 62.

[48] Larrabee, "Guinea Pig Books," 72.

[49] Kallet and Schlink, 6, 14, 303.

[50] Larrabee, "Guinea Pig Books," 72.

same may be said of the guinea pig muckrakers of the 1930s. Yet there were differences between the two groups. For one thing the writers of the New Deal would reach the public less through popular magazines, the medium of their Square Deal counterparts, than through books. One result of this was that the number of people aroused by these writers was far less than those aroused by exposés of Upton Sinclair and Samuel Hopkins Adams. Secondly, while the old muckrakers worked hand in glove with Wiley to gain a new law, those of the 1930s frequently spoke not as collaborators of food and drug officials but as some of their most severe critics. Kallet and Schlink devoted an entire chapter of *100,000,000 Guinea Pigs* to the ergot-ether-digitalis episode and uncritically accepted Howard Ambruster's charges against FDA as true.[51]

There were also similarities between the old and new literary crusaders. A sharp moral fervor permeated the writings of both. The guinea pig writers were concerned with the public's economic loss but the matter was one of more than mere dollars. The enemy was an immoral malignancy eating away at the health of the nation. Kallet and Schlink called the medicine men the "rankest flower in the garden of rugged American individualism." Those who peddled adulterated goods, the 1935 volume *Counterfeit* charged, "are willing to rob him [the consumer] of his health and even of his life if they can make money by doing so." They should be treated accordingly. The food and drug trade argued that demands for regulation proposed in the 1930s were excessive and would seriously injure their industry. M. C. Phillips answered in *Skin Deep* that "if the barn is sufficiently infested with rats . . . perhaps burning down the infested barn is cheaper in the long run than paying doctors' bills, especially when the rats pay no bills."[52]

[51] James Harvey Young, "The 1938 Food, Drug, and Cosmetic Act," in *The Government and the Consumer* (Chicago, 1962), 13-14; for the chapter mentioned, see "Three Drugs and the Law," in Kallet and Schlink.

[52] References are respectively: Kallet and Schlink, 171; Arthur Kallet, *Counterfeit—Not Your Money but What It Buys* (New York

The guinea pig volumes were strikingly like each other. None was scholarly or particularly original. Most drew heavily on the same source material—the files of Consumers' Research and published data from the American Medical Association. Their tone and rhetoric were similar, so much so that whole chapters from most volumes could be interchanged with no break in continuity. Even chapter titles read much the same. In *Not To Be Broadcast*, Ruth Brindze labeled a section on proprietary nostrums, "The Medicine Men's Show." *100,000,000 Guinea Pigs* called a like section "The Quack and the Dead." In *40,000,000 Guinea Pig Children* by Rachel Palmer and Isidore Alpher it was "Don't Let Them Be Poisoned." Almost invariably each book found in the advertising industry the ultimate villain that paved the way for the adulterators.

The guinea pig books lacked depth and generally exaggerated their case. They were unfair to FDA, but the volumes were popular. What was said often angered the personnel of the agency but very early they recognized also a "silver lining." After leafing through a 1932 article by Schlink and Kallet, in the *Nation,* entitled "Eat and Be Poisoned," FDA's Paul Dunbar wrote Walter Campbell, "certainly Schlink had no intention of posing as an ally of the Food and Drug Administration but unconsciously he may be a very valuable one."[53] Out in Minnesota Dr. Walter Alvarez of the Mayo Clinic had the same feeling. He was not pleased at the overly "fussy" and "apprehensive" text of *100,000,000 Guinea Pigs* but still, he wrote Swann Harding of the Agriculture Department, "I think the value of the book lies . . . in its calling attention to the rottenness and inadequacy of our laws."[54]

The new stream of muckraking works reinforced the long-standing critique of national quackery by the Ameri-

1935), 40; Phillips, *Skin Deep: The Truth about Beauty Aids* (New York, 1934), 59.

[53] Dunbar to Campbell, October 8, 1932, Correspondence, Box 17.

[54] Alvarez to Harding, April 29, 1933, Papers of T. Swann Harding, LC 111-23-P, 3, Manuscript Division, Library of Congress, Washington, D.C.

can Medical Association. Indeed the AMA critique was the link between the new muckrakers and the original muckrakers of the Theodore Roosevelt era who had helped to push through the 1906 law, as well as other significant reforms. The most important single figure within the AMA's assault was Arthur J. Cramp, a bitter opponent of proprietary and medicinal abuses and by the early 1900s in charge of AMA activity in the field. In 1911 he compiled *Nostrums and Quackery,* followed by a larger sequel in 1921. It was an 800-page who's who in quackery.[55]

In April 1923, the AMA launched *Hygeia,* a publication designed to give laymen more health information. The editor was Morris Fishbein, himself a dedicated crusader against medicinal abuses and the author of *Fads and Quackery in Healing,* among other works. Fishbein encouraged Cramp to use *Hygeia* as a vehicle to expose the quack to the public and he did so with zeal. Bernarr MacFadden's *Physical Culture* was one of the first targets for the questionable quality of both its articles and its advertising, but the list of targets grew quickly. In the 1920s Cramp came to head the AMA's Bureau of Investigation where he blasted away at medical quackery for the rest of the decade and beyond. Till the arrival of the new muckrakers, however, the AMA fought a lonely battle.[56]

The popular tide of consumer discontent which was beginning to build by the early years of the Great Depression and which sustained the literature of the new muckrakers also showed up in other ways. One was the growth of the highly militant Consumers' Research organization. Founded in 1929, in four years it had grown from less than 1,000 to 45,000 members. Its *General Bulletin* kept consumers acquainted with regulatory activities of the government as well as the quality of many consumer goods on the market. Yet CR's influence and significance far exceeded its actual membership. The organization served as a coordinating body for Congressional and other sym-

[55] Young, *Medical Messiahs,* 131.
[56] *Ibid.*

pathizers of consumer legislation. It provided these people with needed data, and it served to inspire, by means of various publications, the indignation of the public at the state of the drug, cosmetic, food, and advertising market.[57] CR's information was consistently sensational, and it was widely read. "The astounding thing," as a physician wrote Rexford Tugwell in early 1933, "is the receptive attitude of the general public" to that organization.[58] Here again was the symptom of change and perhaps the significance of CR—as a catalytic agent.

It would have been difficult for Walter Campbell to recognize the change in 1933. The alterations in consumer sentiment were too much like Lewis Carroll's Cheshire Cat, only presenting a smile. More prominent was the fact that the nation was immersed in depression, and this economic catastrophe had accelerated the evils with which FDA must cope. Bankrupts from other industries found lucrative possibilities in the ranks of quackery. Sharpened competition promoted more vicious advertising claims. An economically depressed public became more receptive to self-medication to avoid "unnecessary" doctor bills. Americans were spending $350,000,000 annually on patent medicines, enough for three or four bottles for every man, woman, and child in the country.[59]

By November, FDA's 1932 prosecutions numbered 1,307 plus 1,260 seizures of consignments of goods. In December 1932 alone, the agency seized 100 consignments.[60] Whatever the significance of Consumers' Research, there had been no popular ground swell to strengthen the authority of the drug administration. CR might have been pushing for change, but there were many other organizations which should have been fighting but were not. The American

[57] Corbett, "Activities of Consumers' Organizations," 63-64.

[58] Dr. Walter O'Kane to Tugwell, February 23, 1933, Correspondence, Box 17.

[59] Schlink and Kallet, "Poison for Profit," *Nation* 135 (December 21, 1932), 610.

[60] *Press Release*, USDA, November 14, 1932, and January 11, 1933, in Papers of Josiah W. Bailey, Duke University Library, Durham, N.C.

Home Economics Association was one. Too much of their concern was devoted, as a critic stated, "to such burning problems as the 'Factors Controlling Internal Temperatures of Butter Cakes During Baking.'"[61] No, any real change in consumer opinion was too much like the Cheshire Cat as yet.

The Cheshire Cat is an appropriate simile. Walter Campbell might well see his daily labors as carried on in a kind of wonderland. Much of his time was spent in the unreality of panaceas conjured up by commercial advertising. He worked in the sphere of extract of horsetail weed, purported to cure tuberculosis; of "ginger jake" tonic, which paralyzed 150 people in 1931. His world was that of Lash Lure, the eyelash dye which could blind its users, and of Dr. Siegal's "Great Mechanical Developer" for sexually impotent males. Doubtless it did seem a bit unreal. Certainly the AMA's Arthur Cramp felt so. He always kept a copy of Lewis Carroll's *Alice in Wonderland* on his desk and reputedly read a chapter before writing up a case. It put him in the mood.[62]

Ironically, the same image would be utilized by H. B. Thompson, general counsel for the Proprietary Association at Senate Hearings on a new food-drug bill in December 1933. He had located the precedent for the bill, Thompson announced. FDA must have found it in the trial of the Knave for stealing tarts. Not unlike the Queen's demands, the proposed new law's demands called for a verdict and then the trial.[63] The bill was open to criticism, and it did need revision. Yet it is well to remember that in Carroll's tale the only one who found fault with the logic of the Queen was Alice. She could awaken and leave Wonderland. Walter Campbell and his colleagues could not leave. When the drug unit chief set out to see the Assistant Secretary

[61] Corbett, "Activities of Consumers' Organizations," 63-64.
[62] Young, *Medical Messiahs*, 133.
[63] *Food, Drugs, and Cosmetics, Hearings before a Subcommittee,* U.S. Senate, 73rd Cong., 2nd Sess. on S. 1944 (December 1933), testimony by H. B. Thompson, 173.

that wintry morning in 1933, he unknowingly had set out also on a long voyage to make his world more realistic. In June 1938 a new and far more adequate food and drug law would come into being, but that event was five very long years away.

II

"LYDIA PINKHAM
AND OTHER WASHINGTONIANS"

This measure frankly challenges the sacred right of freeborn Americans to advertise and sell horse liniment as a remedy for tuberculosis—or, to phrase it in a wholly different way, his God-given right to advertise and sell extract of horsetail weed as a cure for diabetes. This is precisely the sort of constitutional question which stirs men to the very depths of their pocket-books.

PAUL ANDERSON,
NATION, 1933

WHY NOT write an entirely new law," FDA's Paul Dunbar suggested to his chief over lunch. They were discussing Campbell's morning visit with Tugwell. The new Assistant Secretary had been sympathetic once the conversation got underway. In the end he had even been receptive to some revision of the old 1906 food and drug statute. More surprising, he had followed through. Before noon Tugwell phoned Campbell's office to announce he had gained President Roosevelt's approval for revision. Dunbar was excited. The present law was simply out of date and old-fashioned; amendments would just complicate matters. The thing to do, he urged Campbell, was "to start from scratch."[1]

At the time Campbell was noncommittal. Yet the suggestion did strike a responsive cord. In a very few days things began to move fast—perhaps too fast. Members of the Food and Drug Administration staff were working until two and three in the morning. Law Professors Milton Handler of Columbia University and David Cavers of Duke were

[1] Dunbar, "Memories of Early Days of Federal Food and Drug Law Enforcement," *FDC Law Jnl.* 14 (February 1959), 135.

brought in to assist in the drafting. A new statute was on the way.

These proceedings did not long remain secret. Rumors spread rapidly through the affected industries, and businessmen grew nervous. During the months of April and May 1933, the FDA was flooded with requests about the bill's provisions.[2] When public hearings were finally held on the proposed new statute in late April, they did little to mollify the growing apprehension of the trade, perhaps the reverse. Manufacturers wanted something in black and white. The government presented no text. It was still in the drafting stage, and the conferences were called to get trade suggestions. Maybe so, *but*, mumbled members of the trade.

At the drug hearing on April 27, Tugwell called the meeting to order and then promptly departed, leaving the proceedings to Campbell. This did not seem to indicate the Assistant Secretary's concern for the drug manufacturer's point of view.[3] The government should have had a draft to present. All the concerned manufacturers were pretty sure about that. They tended to take the omission as a sign of distrust.[4] FDA must be withholding the text on purpose. Questionnaires were being sent out to consumers for their opinion of proposed articles. FDA officials were making radio talks on the new bill.[5] Pamphlets with such titles as *Why We Need a New Pure Food Law* were being distributed to the public from FDA field offices over the coun-

[2] Many such letters may be located in FDA General Correspondence, Box 315, Decimal File .062, RG 88, NA.

[3] "New Food and Drug Act," *Drug and Cosmetic Industry* 32 (May 1933), 448. Hereafter cited *D&C Industry*.

[4] *Ibid.*; *New York Journal of Commerce*, April 11, 1933, FDA Scrapbooks, Vol. 1, located in "Records of the Office of the Commissioners of FDA." Hereafter cited as FDA Scrapbooks, Vol. —. "Government Distrust," *D&C Industry* 32 (May 1933), 405; Charles W. Dunn, *The Federal Food, Drug, and Cosmetic Act* (New York, 1938), Appendix A, records of the preliminary conferences, April 1933.

[5] Radio script, "Should the Federal Food and Drug Act Be Expanded?"; Consumer questionnaires, Correspondence, Box 324.

try.[6] It seemed hard to believe that *someone* did not have a precise draft.

The chief concern of the trade was that a new law might be pushed quickly through Congress before they could do anything about it. That was a baseless fear. Even if the Food and Drug Administration was hoping to do this, which they were not, it was impossible. To begin with, Tugwell and Campbell had no Presidential backing for such a move. For that matter, one of the bill's greatest handicaps throughout the five-year fight to gain passage was Roosevelt's lukewarm interest in the project. FDR had done little more in 1933 than concur in Tugwell's suggestion that revision was a good idea. Even at that, as Tugwell later stated, he had spurred the President's interest by playing upon his sense of continuity with Theodore Roosevelt. TR had signed the first food and drug bill and "no President since had dared tighten its restrictions." The sense of continuity, however, was not such as to demand immediate action. FDR made this clear at a May press conference. The idea was simply to get the measure introduced during the spring session of Congress. It was not "emergency legislation" and would wait till fall for any consideration.[7]

Even if the President had favored rapid passage, the likelihood of success would have been slight; the bill lacked strong public support. The opposition was quite correct in their charges that it had been promoted almost totally by enthusiasts in the food and drug unit. Half the battle in the next years would be the difficulty of getting and keeping popular backing for legislation. Given the degree of public apathy, FDA was not likely to score a quick victory over the trade lobbies; the lobbies were too powerful. Experience in the fight for the 1906 statute proved it. Every-

[6] Correspondence, Box 319. FDA files show sundry requests by field offices for pamphlets.

[7] Rexford G. Tugwell, "The Preparation of a President," *Western Political Qtly.* 1 (Winter 1948), 133-34; *Roosevelt Press Conferences*, May 26, 1933, Vol. 1, Franklin D. Roosevelt Papers, Roosevelt Library, Hyde Park, N.Y.

one, including Roosevelt, understood revision would be an uphill fight. The best Roosevelt could say, even to personal correspondents who urged his legislative support, was: "I hope we can get it through in spite of the lobbies."[8]

Further proof of the problems involved in passage was the difficulty FDA found in gaining a Congressional sponsor for the legislation. Both Representative Marvin Jones and Senator Ellison Smith, chairmen of the Agriculture Committees in their respective houses, suddenly found they were too busy to undertake leadership of the measure. They were not the only ones. As one staff member of the Food and Drug Administration wrote a friend sometime later, "Dr. Tugwell and Mr. Campbell literally peddled the thing up and down the halls of Congress."[9] This situation ended only when Senator Royal Copeland came forward and volunteered to take the job. No, it was not very likely the new bill would be passed too quickly.

Copeland got the draft measure in late May and introduced it in the Senate on June 12, as the emergency Congressional session drew to a close. At the time he had not read the measure through. When he finally did so four months later he found that Senate bill 1944 contained drastic changes—some too drastic even for him.[10] It greatly

[8] Roosevelt to Harvey Cushing, April 21, 1933, OF 375, Roosevelt Papers.

[9] Rexford G. Tugwell to Roosevelt, June 1, 1933, *ibid.*; Lamb to James Rorty, April 11, 1934, Correspondence, Box 438.

[10] James Harvey Young, *Medical Messiahs*, 164. Some of these features may well have been the result of "Tugwellian" strategy. The Assistant Secretary quickly sensed the strength and vigor of the coming opposition from the affected industries. While he remembers no details, Tugwell recalled in 1968 that when questions arose during the drafting process over whether to alter aspects of the bill in an effort to conciliate the business community he had always urged that the more "disagreeable" language remain in the text. That way there would be "things" to give up later as apparent concessions to the trade. Tugwell, Rexford G., An Interview with Charles O. Jackson, June 7, 1968. Transcript in Oral History Collection, National Library of Medicine, Bethesda, Maryland. For Copeland's attitude about the measure see *Food, Drugs, and Cosmetics, Hearings before a Subcommittee of the Committee on Commerce, U.S. Senate, 73rd Cong., 2nd Sess. on S. 1944* held December 7 and 8,

expanded the government's control over proprietary remedies. No longer would the drug unit have to prove fraudulent intent to take action against a nostrum. A drug was
misbranded if its labeling made any therapeutic claim,
even by inference, which was contrary to the general agreement of medical opinion.

Labels must specify that the contents were a palliative
and not a cure, when such was the case. The bill called for
label disclosure of all medical ingredients instead of the
mere eleven drugs prescribed in the old law. Many proprietary products which were risky to take, even though the
labels made no false claims, were outlawed. The term
"drug" was broadened to encompass mechanical devices.
Even products which strictly speaking had little to do with
illness, such as weight reducers, were brought under control. Adulteration was defined to include any product dangerous to health when used according to label directions.

The food industry came in also for a share of the new
regulation. Labels must disclose all ingredients in order of
predominance by weight. The government would gain the
right to establish identity standards for quality and fill of
containers. A product was misbranded if it failed to meet
those standards. The definition of adulteration was broadened to apply to products containing poisonous substances
in excess of tolerance levels set by the Secretary of Agriculture. Inspectors were authorized to make checks of es-
tablishments in which food, as well as drugs and cosmetics,
was manufactured or held. Where this "privilege" was
denied, injunctions might be sought by the government to
deny a company the right to engage in interstate shipment
of goods. Provision was further made for a system of voluntary factory inspection services at the owner's expense, but
under such conditions that manufacturers would almost be
forced to accept it.

Cosmetics were brought under the general provisions of
the bill, and advertising felt the weight of new regulations.

1933. Hereafter these Hearings are cited as *Senate Hearings on S.
1944* (December 1933).

The promotion of foods, drugs, and cosmetics must adhere in general to those standards applicable to labeling. In regard to certain conditions where self-treatment was dangerous or futile there was a total prohibition on advertising remedies. Finally, violators of the law were subject to much stiffer penalties than before, in some cases even prison.[11]

The lines of revision under the so-called Tugwell bill were not theoretical propositions but were dictated by twenty-seven years of experience with the old statute. The changes were radical, however; there could be little debate on this point. *Business Week* asserted that the bill sounded like the last chapter of *100,000,000 Guinea Pigs*. There was even some mistaken suspicion that the book's author, F. J. Schlink, had assisted in the drafting process.[12] The trade was shocked. Mildest reaction came from the food industry, and this would remain the case throughout the career of the proposed legislation. They were hardly happy with all the provisions. The drafting was ambiguous and vague. They had reservations on the grading and labeling provisions. Certainly, and upmost, they felt the bill lodged too much discretionary power with the Secretary of Agriculture.[13] Still, as industry spokesman Charles Dunn stated, "Mr. Campbell is undoubtedly right in his major premise . . . that the Federal Food and Drug Act requires revision to cure serious defects in it."[14]

The rhetoric of the food industry was seldom directed at opposition to the bill or revision per se. The most pronounced initial plea of the food men was that their industry

[11] A convenient summary of provisions under S. 1944 may be found in Dunn, *Federal Food, Drug, and Cosmetic Act*; also useful is Young, *Medical Messiahs*, 164-66.

[12] "Prof. Tugwell Offers the Food and Drug Administration an Entire New Set of Teeth," *Business Week*, June 10, 1933, 10.

[13] Laurence V. Burton, "What the Food Manufacturer Thinks of S. 1944," *Law and Contemporary Problems* 1 (December 1933), 120-25.

[14] *Senate Hearings on S. 1944* (December 1933), testimony by Charles W. Dunn, 139.

deserved a separate law.[15] This plea was a bit more than a simple desire for industrial legislative integrity. At least some of those who called for a separate bill believed the revision drive was inspired by the need for stricter regulation of drugs and cosmetics.[16] Assuming that this view was correct, separate legislation might well relieve the food industry of some of the general burdens of the bill which were primarily aimed at the drug trade. Also pertinent to the reaction of the food men was the fact that the new measure effected fewer changes in their industry than in the other concerned trades. Sebastian Mueller of the Heinz Company made this plain at the December hearings. "It is our feeling that the bill as drawn is not impracticable and unreasonable," he stated, "because its enactment . . . would not necessitate any changes in any of our present manufacturing practices."[17]

Food spokesmen had a great deal to write and say. At times FDA was deluged by the unsolicited "helpful" commentaries of people like Charles Dunn, counsel for the Associated Grocery Manufacturers of America. Yet the briefs of these men were almost always sophisticated and respectable, and they demonstrated a willingness to compromise. FDA might tire of Dunn's overactive pen but he was treated with constant respect. This was not the case with the less restrained commentaries from other segments of the affected industries. Drug Administration files disclose more than one brief by Frank Blair, President of the Proprietary Association, passed about the bureau under a large penciled caption which read "More Hokum."[18] Blair and other industry spokesmen were well aware of this difference in treatment and it remained a constant source of irritation to them.[19]

[15] *Ibid.*, testimony by Samuel Fraser, J. W. Herbert, Sebastian Mueller, 205, 425, 136, respectively.
[16] *Ibid.*, testimony by Herbert, 425.
[17] *Ibid.*, testimony by Mueller, 136.
[18] Blair to Walter Campbell, November 15, 1933, Correspondence, Box 324.
[19] Ole Salthe to Royal S. Copeland, April 16, 1935, Correspondence, Box 565.

There were obvious reasons for the contrast in handling. The December public hearings serve as a case in point. While Dunn admitted the need for revision, H. B. Thompson, counsel for the Proprietary Association, compared the bill to parts of *Alice in Wonderland*. He insisted that the only way it could be amended "was to strike out all after the enacting clause."[20] The position of the medicine industry, prescription and proprietary, was keynoted by Dr. James Beal of the National Drug Trade Conference. The Conference was an organization made up of delegates from a number of national pharmaceutical associations representing a variety of interests in drugs. Beal disliked virtually every major feature of Senate bill 1944 from its definition of drugs through the penalties imposed for violations.[21] Other members of the Conference promptly concurred with his views.

A striking feature of the rhetoric was the reverence in which the medicine industry held the 1906 statute. Dr. Beal called it "the most efficient act of its kind in existence." One journal called it a "monument."[22] If there must be change, then by all means let it come as an amendment to the old statute. This approach would preserve years of valuable court precedents. After all, abuses in the market were few in number. An entirely new bill was not warranted—especially such a radical bill. "This would be," said another periodical, "like burning down the house to get rid of the mice."[23] It was all too apparent by December

[20] *Senate Hearings on S. 1944* (December 1933), testimony by Thompson, 172.

[21] *Ibid.*, testimony by Beal, Thompson, Bruce Kamer, 83-120, 172, 391, respectively. While the Drug Trade Conference included both ethical and proprietary manufacturers it would appear that Beal's comments more clearly reflected the grievances of the latter rather than the former. Prescription manufacturers very early showed ambivalence on the matter of a new law.

[22] *Ibid.*, testimony by Beal, 84-85; *Standard Remedies* 20 (November 1933), 5.

[23] "The Case against the Tugwell Bill," *Advertising and Selling* 22 (November 9, 1933), 13-14.

1933 that much of the trade was, in the words of *Printers'
Ink,* out to "Beat the Tugwell Bill."[24]

In this attack the affected industries were prepared to
use every available tack and tactic. Perhaps there was reverence for the 1906 law now, but there was much irony,
even about the call for its preservation. As Charles LaWall,
a co-worker with Harvey Wiley on the first drug bill,
pointed out in 1937, by which time opposition to a new
statute was considerably diluted, "the opponents of this bill
have made every one of the specific objections to its passage that were made . . . of the original Hepburn Bill in
1906."[25] In 1933 the trade had *some* objections to specific
clauses but more characteristic was a generalized assault
on the motives and philosophy behind S. 1944. Specific objections came later and only with the realization that some
bill would pass into law.

One tack was to deny the serious inadequacy of the old
Wiley law. Everyone knew, the opposition stressed, there
had been no public clamor for new legislation. The bill
"is simply something that Mr. Campbell and his staff wants
done." There was even some doubt, G. A. Nichols pointed
out in *Printers' Ink,* as to whether the proposed statute was
an Administration measure.[26] On little more than a whim
FDA was out to destroy not only valuable legal precedents
but also the years of work it had taken to bring state laws
into reasonable conformity with the federal law.[27] The food
and drug unit wanted simply to get powers the courts had
quite correctly denied them in the past, charged *Standard
Remedies,* a voice of the major proprietary drug manufacturers.[28] If the bill did pass, it would not have popular

[24] G. A. Nichols, "Beat the Tugwell Bill," *PI* 165 (November 2,
1933), 6ff.

[25] LaWall, "Fads and Frauds in Foods and Drugs," *American
Journal of Pharmacy* 109 (March 1937), 119. Hereafter cited *AJP.*

[26] Nichols, "Beat the Tugwell Bill," 10-11; Frank Blair, "Effects
of Proposed Legislative Changes upon Industry," *Standard Remedies*
20 (November 1933), 6.

[27] *Ibid.,* 5.

[28] "Food and Drug Bills Pending," *Standard Remedies* 20 (September 1933), 5.

support and would prove unenforceable. It would be another Volstead Act and provide the same kind of experience.[29]

The proposed new law was bad, if for no other reason than timing. The nation was laboring under the weight of the Great Depression. In such perilous times recovery must be the primary concern. The Tugwell bill hardly promoted recovery. Ray C. Schlotterer of the New York Board of Trade pointed out to the Senate Commerce Committee that the measure affected the employment of approximately 1,775,000 people as well as products in the value of $17 billion. Such drastic and arbitrary control of the affected industries, as proposed in S. 1944, could only be destructive. In short, Schlotterer charged, the bill was anti-NRA and defeating of its self-regulatory principles.[30] NRA Administrator Hugh Johnson was besieged with like assertions through the mail. Commercial concerns received similar material. Typical were the posters from the New York Board of Trade bearing the caption, "The Tugwell Bill Is Anti-NRA," followed by an elaborate diagram to prove the measure's ill effects. Trade journals echoed the cry.[31]

But the ills of S. 1944 were more than timing. The measure represented a radical shift to government by bureaucratic dictum. According to *Drug Trade News* the bill was "un-American," "discriminatory," and "insidiously selfish."[32] For those who feared the dangers of such a shift, Chester Wright, pontificating in *Printers' Ink*, suggested that ample ammunition was to be found in close scrutiny of a book by Lord Hewart of Bury, Lord Chief Justice of England, written some years earlier. Titled *The New Despotism*, it provided a picture of government by bureaucracy which could only be described as "shocking." Hewart found the seeds of

[29] *Milwaukee Sentinel*, October 16, 1933, FDA Scrapbooks, Vol. 1.
[30] *Senate Hearings on S. 1944* (December 1933), testimony by Schlotterer, 278.
[31] *OP&D Reporter*, October 23, 1933; *Drug Trade News*, October 30, 1933, FDA Scrapbooks, Vol. 1.
[32] *Ibid.*; *San Francisco Chronicle*, December 14, 1933, FDA Scrapbooks, Vol. 2.

this situation in skeleton legislation where large grants of discretionary power was conveyed to bureaucrats—according to Wright, "a perfect description of the Tugwell bill."[33] "Unlimited power entrusted to bureaucrats warps their judgment on the opinions they might have as normal citizens," warned William L. Daley of the National Editorial Association.[34]

The discretionary power of S. 1944 was a fearful thing. Lawyers for *Standard Remedies* informed the journal that if the bill passed "no manufacturer can possibly continue in business except by the grace of the officials in Washington."[35] The Department of Agriculture would become a "virtual dictatorship over the trade." Bureaucrats would be able to decide purely on the basis of their whims what advertising copy was acceptable. Powers to establish government grading of goods would, in effect, mean that the Secretary of Agriculture could determine which companies survived. The failure to receive government sanction would be tantamount to denunciation of the product.[36] Remember that some dermatologists even believe such harmless substances as talcum, starch, and fatty oils are injurious to the user, Dr. Beal of the National Drug Trade Conference warned. Some day there could be people in FDA who would subscribe to that idea and exercise the discretionary power of the bill to that end.[37]

There was some validity in the charge that S. 1944 allowed too much discretionary power. The bill had been hurriedly drawn and its provisions were too loose. Certainly Royal Copeland admitted that some tightening up and clarification of provisions were necessary. The surpris-

[33] Wright, "Despotism and the Bureaucrats," *PI* 165 (December 21, 1933), 60.
[34] *Senate Hearings on S. 1944* (December 1933), testimony by Daley, 336-37.
[35] *Standard Remedies* 20 (August 1933), 1.
[36] *Drug Trade News*, September 18, 1933, FDA Scrapbooks, Vol. 1; Wright, "Tugwell Bill Would End Value of Brands and Trade-Marks," *PI* 165 (December 14, 1933), 6, 10.
[37] *Senate Hearings on S. 1944* (December 1933), testimony by James Beal, 91.

ing thing was the extent to which the trade would go to make the point. Reaching hard, they even drew on their archenemy Arthur Kallet, co-author of *100,000,000 Guinea Pigs*. A trade-lobby pamphlet used a recent article by Kallet in *Tide* magazine as its text. He had stated that the passage of the Tugwell bill would actually invite vitiation by the pressure groups it intended to control. The Secretary of Agriculture would have so much discretionary power that he would be constantly subject to great pressure from the concerned industries.[38]

Trade alarm in this matter was real. Even more, the opposition feared the person of Rexford Tugwell. One of the worst obstacles that the proposed drug bill would have to overcome was the mere designation "Tugwell bill." Opponents were always clear about this reservation. As Laurence Burton of *Food Industries* explained, "some of these fears would never have come to light had not one of the authors of the bill . . . , Prof. R. G. Tugwell, revealed so much of his economic philosophy in various publications and public utterances."[39] The handsome, debonair Assistant Secretary was a liberal of the far-left type and, what's more, made no secret of the fact.[40] Insofar as the trade was concerned, he was *at best*, said *Printers' Ink*, "a theorist of the dilettante variety who is not altogether sure what he is talking about. He sincerely wants to accomplish some good for humanity. . . . But his best, in this case, is pretty terrible."[41]

At worst, and for too many members of the affected industries, he was little short of an agent for the Third Internationale. "The world knows that he has visited Russia," as one writer put it, "and has found its institutions acceptable, that apparently he believes that packaging and advertising constitutes economic waste that should be pre-

[38] Pamphlet, *Remarks on the Tugwell Bill*, Bailey Papers.
[39] Burton, "What the Food Manufacturer Thinks," 121.
[40] For information on the relationship of Tugwell to New Deal policy see Bernard Sternsher, *Rexford Tugwell and the New Deal* (New Brunswick, N.J., 1964).
[41] Nichols, "Beat the Tugwell Bill," 6.

[35]

vented."[42] Much of the feeling against Tugwell sprang from exaggeration of his sentiments but he did believe that there was much waste in the American economic system. In his 1933 work, *The Industrial Discipline,* he had stated that nine-tenths of advertising served no useful purpose.[43] The philosophy of the volume also made clear, so far as the National Association of Retail Druggists was concerned, that Tugwell was no friend of package or proprietary medicines.[44]

"Tugwellomania" became an industrial disease. David Cavers, who helped draft S. 1944, characterized its symptoms. They were "cold feet, red spots before the eyes, a loss of the sense of proportion and delusions of persecution. During an attack the patient will try to wreck all progressive legislation with which he comes in contact."[45] Tugwell became a legislative whipping boy. Many people saw this, including Senator Copeland. How interesting it was, he commented to FDA's Paul Dunbar, that whenever people spoke badly of the proposed measure, it was called the "Tugwell bill." If they mentioned good points they named it the "Copeland bill."[46] Even the press on occasion found "Tugwellomania" too much to accept. Heed the ominous warnings of "Today's Hero," Acidosis B. Doakes, lampooned one journal:

> Mister Tugwell is a theorist who not only wants to eliminate rugged individualism . . . but he is a dangerous dreamer. I happen to know a school teacher in Gary, Indiana, who overheard a conversation in a Washington cafeteria, in which it was predicted that the Brain Trust intends to make the manufacturer of aspirin tablets a government monopoly. In fact the plan is to abandon

[42] Burton, "What the Food Manufacturer Thinks," 121.
[43] *OP&D Reporter,* October 9, 1933; *Natl. Assn. of Retail Druggists Jnl.,* October 19, 1933, FDA Scrapbooks, Vol. 1.
[44] *Ibid.*
[45] Cavers, "Tugwellomania," *Food Industries* 9 (January 1934), 2.
[46] Dunbar, "Memories of Early Days," 136.

both the silver and gold standards . . . and issue aspirin tablets from the Federal mint.[47]

Yet the problem was not even so simple as the left-wing inclinations of Rexford Tugwell. There were other reasons for the affected trades to fear. Dr. D. Aitchison of the so-called National Liberties Association pointed out the additional danger. He could not imagine a Congressional committee giving sanction to S. 1944 *but,* "I can see how a man who is already linked up with the medical trust that he would advocate anything to further the ends of the trust that is greasing his palm with blood soaked dollars."[48] The American Medical Association had nothing to do with the drafting of the bill, and was at best apathetic to S. 1944. Whatever the case, the AMA came off as an archvillain in the propaganda of the proprietary medicine trade.

There were far too many people like Morris Bealle, editor of *Plain Talk,* who were convinced that even Tugwell had been duped by the "dictators" of the AMA and that the bill was aimed at abridging the "sacred right" of self-medication.[49] The advertising provisions alone would take away from the public that knowledge needed to handle simple ailments. People would have to visit a physician to get medicine they could otherwise purchase, without a professional fee, at the local drugstore.[50]

It is difficult to believe that many manufacturers or advertisers really saw the imminent collapse of the package

[47] "Today's Hero," [journal not indicated], Fall 1933, FDA Scrapbooks, Vol. 1.
[48] *Senate Hearings on S. 1944* (December 1933), testimony by Aitchison, 132-33. Aitchison did not reveal the nature of the organization's membership. He simply stated that he represented 40,000 Americans.
[49] Bealle to Tugwell, November 21, 1933, Correspondence, Box 430. See also *Drug Trade News,* September 18, 1933, FDA Scrapbooks, Vol. 1; "The Drug Act," *D&C Industry* 33 (September 1933), 223; Jonathan Mitchell, "Heap Bad Medicine," *New Republic* 76 (November 8, 1933), 353.
[50] Herbert Mayes, "The Tugwell Bill Can Give the Drug Business to Physicians," *Am. Druggist* 138 (November 1933), 32; "The Food and Drug Act," *D&C Industry* 33 (September 1933), 225.

medicine industry as a realistic consequence of the Tug-well bill. It is equally difficult to believe that involved tradesmen actually believed the bill to be a plot of the AMA, but the charges were effective propaganda. Letters such as the one received by Senator Bailey from Emma Carlisle of Whitaker, North Carolina, testify to the power of the message. She had just read a newspaper article about the bill, which she enclosed. In a badly written scrawl she pitifully inquired, "If any one has sick headache would it be a violation of the law to make a cup of thyme tea for relief? The poor can't have a Doctor for every minor scratch."[51]

The force and organized strength of the opposition were apparently a surprise to FDA. They had expected resistance but clearly had underestimated its power. The opposition not only made their case with sensational vigor but organized into effective lobbying units with exceptional speed. Nor was that organization achieved without over-coming serious obstacles. There was significant division within the ranks of the affected industries. The Association of National Advertisers found its 1933 convention totally unable to produce a formal statement of position on S. 1944. One big consideration was that the food manufacturers hesitated to align themselves with the drugs and cosmetics manufacturers.[52] Even within the drug industry there was division. The fact was that some segments of the industry might benefit from passage at the expense of others. Thus many producers of prescription products assumed their volume of business would increase if proprietary advertising were curtailed.[53]

The mobilization did take place, however, and on a wide front. Many existing trade bodies were turned almost immediately into vehicles of resistance. Especially militant were the Proprietary Association and the United Medicine

[51] Carlisle to Bailey, December 5, 1933, Bailey Papers.
[52] "Administration Dodges Tugwell Bill Defense," *PI* 165 (November 16, 1933), 44.
[53] "A Divided Industry," *D&C Industry* 33 (October 1933), 327.

Manufacturers of America. Other groups such as the National Drug Trade Conference, American Newspaper Publishers Association, National Association of Retail Druggists, and the National Publishers Association quickly joined the battle. Common weapons used by such organizations were the sponsorship of public protest meetings, radio propaganda, and memorials pressed upon government officials.

Besides the standing groups, the months after the introduction saw a proliferation of ad hoc organizations called into being to join the fight. The largest was the Drug Institute of America, founded by Charles Walgreen of the Walgreen Drugstore chain. Ostensibly it was created to protect "reputable" dealers against "cut-rate" outlets but, in fact, its main concern was opposition to both the NRA program and the Tugwell measure.[54] Other groups whose real purpose was not clearly avowed were the so-called Minute Men, founded specifically to fight drug law revision, along with the later Joint Committee for Sound and Democratic Consumer Legislation and the National Advisory Council of Consumers and Producers.[55]

In the long fight for the 1906 Wiley law the government had gotten much support from popular magazines and even newspapers. The early career of the Tugwell bill suggested this would not be the case again. At least one contemporary saw S. 1944 as the first break between the Roosevelt administration and the press.[56] Since the new

[54] Kenneth G. Crawford, *The Pressure Boys* (New York, 1939), 77; "Drug Institute Makes Its Bows," *D&C Industry* 32 (June 1933), 506.

[55] Sternsher, 231; W. H. Hartigan to George Larrick, November 29, 1933, Correspondence, Box 313. The Minute Men was founded in 1933. Its membership was made up of individuals, as opposed to organizations, connected mainly with the proprietary drug industry. The Joint Committee for Sound and Democratic Consumer Legislation included representatives from particular companies as well as trade associations in the food, drug, cosmetics, and advertising community. The National Council of Consumers and Producers was similar in membership to the Joint Committee.

[56] A case in point is *Ladies' Home Jnl.* See also *Good Housekeeping*: Dr. Walter Eddy, "A New Pure Food Bill" 97 (October 1933),

measure sought to extend FDA control to advertising, some journals which had fought for reform in 1906 were hostile in 1933. In the case of one magazine, *Good Housekeeping*, an initial editorial in October backed the drug bill but a complete reversal of position took place by the December issue. Many observers saw the shift as a direct result of pressure from the business office.[57]

There is difficulty in assessing the actual importance of the advertising dollar in creating journalistic opposition to S. 1944. Elements of the affected industries, particularly the medicine proprietors, considered it a very powerful tool. Some concerns, such as the Creomulsion Company of Atlanta, were extremely blunt in the use of the weapon. That enterprise circularized editors of rural papers with the explicit warning that "if this bill should become law, we will be forced to cancel immediately every line of Creomulsion advertising." Many other companies would presumably follow suit.[58] Even the relatively conservative Associated Grocery Manufacturers of America voted to wire all newspapers with correspondents in Washington that passage would mean a significant loss of advertising revenue.[59]

The so-called red clause was reintroduced as a weapon from the pre-1906 days. This was a clause which provided for the immediate cancellation of advertising contracts in the event of the passage of new legislation.[60] The Proprietary Association even set up a publicity department to turn out, for the beneficiaries of its advertising, canned editorials against the Tugwell bill. For whatever reasons, these

---

94, 96 as contrasted with a later issue, Ernest Calkins, "Another Look at the Pure Food Bill," *ibid.* (December 1933), 90ff. For comment on FDR and the press see George Seldes, *You Can't Do That* (New York, 1938), 91-92.

[57] Ruth Lamb, *American Chamber of Horrors*, 180.

[58] Creomulsion Company to "Gentlemen," October 30, 1933, Commissioners' File, Box 6.

[59] *Food, Drugs, and Cosmetics, Hearings before the Committee on Commerce*, U.S. Senate, 73rd Cong., 2nd Sess. on S. 2800 held February 27 to March 3, 1934, testimony by James Rorty, 411-12.

[60] Lamb, 202-203; Crawford, 79.

editorials showed up in such geographically diverse papers as the *New York Journal of Commerce*, the *Atlanta Journal*, and the *Houston Post*.[61] In fact, only three large metropolitan dailies consistently backed the new food bill.[62] The embattled proponents of drug law revision in FDA believed they were being subjected to a news blackout.

The belief may have been warranted. It was equally likely, however, that the slim coverage was because editors, like the broader public, were simply more interested in other issues. On the other hand, there was certainly reason to question whether the press coverage received was fair. A survey of 104 Washington correspondents indicated room for doubt. A total of 46.2 percent of those questioned believed their journals had been unfair; an additional 32 percent were uncertain.[63]

Whether from indifference or malice, the lack of a positive press was a fact to the drug administration. Many people worked hard to make it otherwise. Campbell, Tugwell, and Cavers, among others, devoted much time to defending S. 1944 in trade periodicals.[64] Writer Swann Harding of the Agriculture Department's information office was encouraged by superiors to lend his pen to the cause. He published often, and through his personal journalistic contacts did yeoman service in the name of food, drug, and cosmetic revision.[65] FDA's information officer Ruth Lamb continuously prompted her contacts, such as Dr. Barbara Beattie, to join in the effort. Not all the at-

[61] Crawford, 79.

[62] James Burrow, *AMA: Voice of American Medicine* (Baltimore, 1963), 278.

[63] George Seldes, *Lords of the Press* (New York, 1938), 300.

[64] Typical examples are: David F. Cavers, "How Tugwell Bill Should Define False Advertising," *PI* 165 (December 21, 1933), 68; Rexford G. Tugwell, *Broadcasting and Broadcasting Advertising*, September 15, 1933, FDA Scrapbooks, Vol. 1; Walter Campbell, "Here is the Administration's Position on the Copeland Bill," *PI* 165 (November 30, 1933), 6ff.

[65] Solon Barber to Harding, June 1, 1933, Correspondence, Box 324. Typical examples are: Harding, "Outwitting the Dogs of Fraud," *Commonweal* 19 (November 24, 1933), 93-95; and "False and Fraudulent," *North Am. Review* 236 (November 1933), 439-47.

tempts paid off. Dr. Beattie's initial editorial endeavors were flatly rejected by *Good Housekeeping*, a fact she ascribed to the "Almighty Advertiser."[66]

Yet there were also notable successes. One was Paul Anderson of the *St. Louis Post-Dispatch*, who had been a friend of FDA at least since he covered the ergot investigation of 1930. He proved very receptive to Miss Lamb's call for aid, worked closely with her, and in the coming years provided extremely strong editorial support in both the *Post-Dispatch* and the *Nation*. For him the issue was simple. "This measure frankly challenges the sacred right . . . to advertise and sell horse liniment as a remedy for tuberculosis," he wrote of S. 1944.[67] Miss Lamb was always delighted with him, and certainly her superiors must have been equally delighted with Miss Lamb. With tireless energy she kept a constant flow of "source" material moving out to periodicals over the country. Her quest for food bill coverage went far beyond the large publications. There was a seemingly endless list of minor journals from *How-to-Eat* to *Needlecraft* to *Mother's Home Life*.[68]

The sheer volume of her work was bound to bring some return and so it did. There were always periodicals which would provide a positive editorial.[69] There were always magazines which went further and told shocking stories of blindness from the use of Lash Lure, an eyelash dye, and the fraudulent dangers of Healthagain, the diabetic "cure" made of epsom salts, cane sugar, alcohol, and spices.[70] Yet the volume of return must surely have been disappointing to Miss Lamb. The bad reviews always outnumbered the good. Reports of journalistic criticism continuously flowed

[66] Beattie to Lamb, November 24, 1933, Correspondence, Box 324.

[67] Anderson, "Lydia Pinkham and Other Washingtonians," *Nation* 137 (December 20, 1933), 699.

[68] The records show dozens of letters to publications. See Correspondence, Box 324.

[69] "The Revolt of the Guinea Pigs," *Christian Century* 50 (November 29, 1933), 1492-93; "Opposition to the Tugwell Bill," *ibid.*, (December 27, 1933), 1628-29.

[70] *Independent Woman*, December 1933, FDA Scrapbooks, Vol. 2.

into Washington from FDA field offices. Perhaps the drug administration staff could not be blamed for a slight touch of paranoia. They were working too hard; the "business office" must be to blame. Walter Campbell was sure of it. The lack of journal support, he wrote District and Station Chiefs, "serves to emphasize the fact that our most fruitful approach . . . is to go direct to the people of the nation through the medium of their clubs and other organizations."[71]

For many years Campbell's organization had assumed it bore a responsibility to inform the public of agency functions through speaking engagements and limited literature. With an eye to the lack of press coverage for food law revision these activities were greatly accelerated after the introduction of S. 1944. Through radio spots, mimeographed material circulated from field offices, and direct mailing of reprinted articles by FDA staff members, the agency began a carefully coordinated effort to build support for the new bill. The practices most infuriating to the trade were speaking engagements—"soapboxing," the industry called it, a "frenzied campaign" among organizations over the country.[72] The reason for the anger was simple. These "talks" were effective, especially those before women's organizations. "Did you ever try to alter the mind of a mother concerning what she regards as a hazard to the life or health of her children?" lamented *Printers' Ink*.[73] By late November field office reports indicated that many offices were providing four to five speakers a week. FDA discovered also that radio could be an effective weapon. Using such vehicles as the National Farm and Home Hour, Copeland, Campbell, and other agency officials pleaded their case over the airways several times in the fall of 1933.

By fall, FDA's most effective single weapon, around

[71] Campbell to all District and Station Chiefs, October 31, 1933, Correspondence, Box 324.

[72] G. A. Nichols, "Food and Drug Bill Revision Sure," *PI* 165 (November 23, 1933), 6ff; Nichols, "Beat the Tugwell Bill," 6ff.

[73] Wayne Calhoun, "Hoi Polloi and Tugwell," *PI* 165 (December 28, 1933), 37.

which many of the other propaganda practices centered, was a series of posters graphically illustrating the inadequacy of the 1906 law. Soon labeled by the press as the "Chamber of Horrors," the series drew heavy crowds at the Century of Progress Exposition in Chicago and was subsequently made available to the public by means of full-size reproductions and photo albums at almost every drug administration field office. The display was put on the road with drug unit inspectors and was loaned out to any organization which requested it.[74] Neatly arranged for all to see were the most worthless, the most fraudulent, and the most dangerous products on the market, from Crazy Crystals to Marmola. Their failings were clearly spelled out. There were Koremlu and Radithor along with a series of cosmetics whose lead base could be harmful to the user. There were the food packages misleading as to their fill. There was a place also for Banbar, the diabetes "cure" made from extract of horsetail weed. Next to the Banbar testimonials were corresponding death certificates.[75]

The trade was livid. This was "dirty business," growled *Drug Trade News*. These tactics were proof that FDA should not get the new power it "craves."[76] The time has come to "fight fire with fire," cried *Drug and Cosmetic Industry*—expose to the public the "therapeutic nihilism" of the drug administration as well as their connections with the "medical trust."[77] Yes, the "Horrors" exhibit was effective. The popular press admitted it. Even Eleanor Roosevelt was shocked by the exhibits and became an ardent backer of a new bill.[78] FDA's Charles Crawford was

[74] "Revision of the Federal Food and Drugs Act," *Jnl. of Home Economics* 25 (November 1933), 780; Young, *Medical Messiahs*, 169-70; E. C. Boudreaux to Washington, D.C., office FDA, September 29, 1933, Correspondence, Box 313.

[75] Larrick to all FDA District and Station Chiefs, September 8, 1933; also *Boston Evening Transcript*, September 26, 1933, and other papers, FDA Scrapbooks, Vol. 1.

[76] October 30, 1933, *ibid.*

[77] "Why Not Tell," *D&C Industry* 33 (November 1933), 431.

[78] *Labor: A National Weekly*, October 31, 1933, FDA Scrapbooks, Vol. 1.

jubilant and highly optimistic. "The favorable reaction we are getting . . . ," he wrote David Cavers, "has surpassed all expectations. I am beginning to wonder if we have not overlooked a bet in not writing into it a more adequate provision to control patent medicine abuses."[79]

Crawford's optimism was unrealistic. What FDA needed and did not have was *organized* support which could help mobilize public sentiment. Thus, within medical circles there were many individuals who supported drug law revision, but the American Medical Association seemed reluctant to align itself with any specific bill. They had taken no part, trade rhetoric to the contrary, in the drafting of S. 1944. William Woodward of the AMA talked privately with Campbell about the bill at least once, and the organization's trustees endorsed the "principles" of revision.[80] That was not, however, the same as active support of the "Tugwell" measure. The AMA did not even send a representative to the December public hearings. They filed a written brief but it was less than enthusiastic about many provisions and noncommittal on others.[81] Some members of the FDA staff keenly resented this lack of positive backing.[82]

Then there was the ever militant Consumers' Research organization which could have served a very useful function in evoking public support for the bill. CR was for revision. The organization even boasted in their *General Bulletin* that they were "responsible for the present wide public agitation for" a new act.[83] Yet, ironically, at this crucial moment in 1933 Consumers' Research proved to be at least as much an enemy to FDA as was industry. In

[79] Crawford to Cavers, September 13, 1933, Commissioners' File, Box 12.

[80] "New Food and Drugs Legislation," *JAMA* 101 (December 9, 1933), 1882-83; telegram, Woodward to Campbell, October 27, 1933, Correspondence, Box 313.

[81] *Senate Hearings on S. 1944* (December 1933), AMA Brief, 461-65.

[82] Lamb to Mary Ross, June 12, 1936, Correspondence, Box 323.

[83] July 1933, FDA Scrapbooks, Vol. 1.

100,000,000 Guinea Pigs CR's Kallet and Schlink had accused the drug unit of lax enforcement and of succumbing to commercial pressure.[84] Now they criticized the "weak" provisions of S. 1944. Portions of it, they charged, were equivalent to taking "a few cartridges" from an habitual criminal, while leaving him his gun and his liberty.[85] They cast serious doubt on Copeland's value as a Congressional sponsor by publicizing his commercial dealings with Fleischmann's Yeast and other companies who would come under the regulations of the bill.[86] They circulated petitions calling for the shift of food and drugs control from the Department of Agriculture to the Public Health Service.[87] Such activities hardly aided the passage of the "Tugwell" bill.

What FDA did have was the support of the national women's organizations, and in the end this aid became a decisive factor in the passage of a new drugs measure. By the December 1933 public hearings the American Association of University Women, the American Home Economics Association, the National Congress of the Parent-Teachers Association, among other such bodies, had endorsed S. 1944.[88] The number of women's organizations would grow in the future. But the days of real assistance were yet to come; the mobilization of the women had just begun by the fall of 1933. For the moment their support was in name only. This was no match for the National Drug Trade Conference.

The lack of organizational support became all the more

[84] J. J. Collins to Henry Wallace, June 1, 1933, Correspondence, Box 3121.

[85] Arthur Kallet, "A Consumer Looks at the Food and Drugs Bill," *Law and Contemporary Problems* 1 (December 1933), 127.

[86] *Senate Hearings on S. 1944* (December 1933), testimony by Kallet, 355.

[87] Mrs. F. E. Malick to Roosevelt, December 19, 1933, Correspondence, Box 314.

[88] Mrs. Harvey W. Wiley to Paul B. Dunbar, December 18, 1933; Mary Smith to Harriet Howe, December 14, 1933; Larrick to FDA District and Station Chiefs, September 26, 1933; Correspondence, Box 320; *Senate Hearings on S. 1944* (December 1933), testimony by Alice Edwards, 345.

important by the late fall when FDA's own propaganda efforts were seriously curtailed. If those efforts had not been so effective the opposition to statutory revision would have been perhaps far less shocked by "this ballyhoo" carried out with the "unlawful" use of public funds.[89] As it was, the program had to be stopped and the opposition began a concerted effort toward that end. Congressmen and other federal officials were deluged with charges from trade spokesmen that the Food and Drug Administration's publicity costs were a violation of the 1919 Deficiency Appropriations Act. This statute prohibited lobbying expenditures by federal agencies.[90] Even men such as Representative John Cochran of the House Committee on Expenditures, who favored S. 1944, became disturbed by complaints. He wrote Secretary of Agriculture Henry Wallace that he doubted "the wisdom of the policy of the Department in going to the people in support of any measure that was in Congress."[91]

Tugwell defended the FDA "educational" program but he too was bothered. He went in turn for legal advice to the Solicitor, who also had qualms.[92] The upshot was that Walter Campbell ordered a complete halt to the publicity efforts, effective on the day S. 1944 was taken up for public hearing. Speaking engagements scheduled after December 7 were cancelled. From mid-November on, the staff of FDA was directed not to take further initiative in loaning out the "Horrors" exhibits, and radio comment on the bill by administration personnel was to cease. Even incoming letters asking how to assist in backing the bill were to be answered by reference to the Deficiency Appropriations Act. In one instance a field office was directed to remove a

[89] United Medicine Manufacturers of America to ——, November 3, 1933, Commissioners' File, Box 10.

[90] A typical example is David Lundy to Senator Bailey, December 29, 1933, Bailey Papers.

[91] Cochran to Wallace, November 15, 1933, Correspondence, Box 313.

[92] Tugwell to Cochran, November 29, 1933; Campbell to all FDA District Chiefs, November 14, 1933, *ibid.*

post office placard which merely directed the public to the location of the "Horrors" display.[93] In later days some of the drug unit's "educational" work would be renewed but never again with the volume or force of 1933. The trade had scored a significant victory.

This was the first of many such victories—the fruits of a determined opposition. The array was imposing. "Here we find them, one and all," Paul Anderson wrote of the Washington scene, "prepared to die for good old Lydia Pinkham, Cascarets, Listerine, and other celebrated benefactors of suffering humanity. This is precisely the sort of constitutional question which stirs men to the very depths of their pocket-books."[94] Insiders such as the *Kiplinger Agricultural Letter* predicted that passage was 90 percent certain for the winter of 1934, but the earlier unbridled optimism of Charles Crawford was now gone from the staff of FDA.[95] They hoped for quick passage. They really hoped hard, but they no longer believed. Even before the December hearings Walter Campbell wrote privately to David Cavers that "we are convinced that the bill cannot be passed without change."[96] The realistic question was what changes and how much.

[93] Lamb to Thelma Murphy, December 20, 1933; Dunbar to all FDA District and Station Chiefs, November 29, 1933; Wendell Vincent to Campbell, December 2, 1933; Leonard Feldstern to Chief, FDA, December 2, 1933. All located in Correspondence, Boxes 319, 323.

[94] Anderson, "Lydia Pinkham and Other Washingtonians," 699.

[95] October 7, 1933, Correspondence, Box 324.

[96] Campbell to Cavers, November 27, 1933, Correspondence, Box 313.

IIII

THE TROUBLES I'VE SEEN

"Keep up the fight, says Mr. ———
Keep up the fight, I say.
At fifty thousand plunks a year
I'll fight till Judgement Day."

> Col. Edmund Taylor in response to 1905
> urgings by counsel for the National
> Wholesale Liquor Dealers for continued
> opposition to the Wiley bill.

RECOLLECTION BY R. M. ALLEN,
MAY 1934

ON February 19, 1934, a harried Royal Copeland rose in the Senate of the United States to introduce S. 2800. This bill, he stated, was "a rewriting of the more or less famous —or infamous—food and drug bill." "Mr. President," he told his colleagues, "I desire to say for myself that I thought I had had all the troubles one could have in this life; but in all my experience I have never had so many worries and so much trouble as I have had in connection with this bill."[1] The feeling was understandable. When he had scrutinized S. 1944 closely in October he became convinced there was need for revision, and expressed this sentiment to the trade at the December hearings.

Copeland labored much of December to draft a more acceptable measure and since Christmas had talked "almost constantly" with representatives of the concerned industries.[2] On January 4, the Senator introduced S. 2000, his substitute for S. 1944. Trade opposition continued unabated. Drug man Lee Bristol, speaking for the Joint Committee for

---

[1] *Congressional Record,* 73rd Cong., 2nd Sess. (February 19, 1934), 2728.
[2] *Ibid.*

Sound and Democratic Consumer Legislation, told the press that while the bill had a "worthy purpose" it still had many elements dangerous to the public interest and it threatened the principles of democratic government. Critical literature poured out to the public from the offices of the Joint Committee.[3] Bristol argued that there was no need to rush any bill through. No emergency situation existed. Edgar Kobak of the Advertising Federation of America felt the same and said this to the public over national radio in January.[4]

Meanwhile the NRA was deluged with letters charging that Copeland's bill violated the principles of the "Blue Eagle."[5] An organized petition campaign against S. 2000 got underway. Professionally printed petition blanks as well as blanks clipped from newspapers arrived at the capital in large numbers. The President and members of the Senate Commerce Committee were priority recipients. This literature charged that the substitute "will set up a virtual dictatorship in the methods of healing."[6] One petition from a Seattle body, called the American League of Medical Freedom, insisted that the only beneficiaries of such a new law would be the American Medical Association.[7]

Senators could hardly ignore the volume of the assault. Under Copeland's leadership S. 2000 was revised in committee. This move caused confusion as to whether current criticism reaching the committee members and other federal officials was intended to apply to the modified version of the bill or the original. To eliminate this problem Copeland chose to introduce the reprint with a different bill

[3] *Drug Trade News*, February 19, 1934, FDA Scrapbooks, Vol. 3; "Copeland Bill Revised Once More," *PI* 166 (February 8, 1934), 18; "The Week," *New Republic* 78 (February 14, 1934), 3.

[4] *Drug Trade News*, February 19, 1934, FDA Scrapbooks, Vol. 3; and January 22, 1934, *ibid.*, Vol. 4.

[5] James Rorty, "Who's Who in the Drug Lobby," *Nation* 138 (February 21, 1934), 214.

[6] Many such petitions are located in Petitions and Memorials File, 73-AJ12, Committee on Commerce, U.S. Senate, 73rd Cong., NA; Correspondence, Box 429.

[7] Petitions and Memorials File, 73-AJ12.

number. There would be public hearings on February 27, he told his colleagues, "when anybody who is interested in firing a few more shots . . . may have that opportunity."[8] For all the concessions in S. 2800 a sizable body of marksmen stood waiting.

The most important changes from S. 1944 were four. The "Tugwell" bill called for full disclosure of formulas on proprietary drug labels. In S. 2800 only specified ingredients needed to be listed. In the new measure publishers would no longer be legally liable for false advertising which they might accept. They must simply turn over to the Agriculture Department the names of manufacturers who prepared the copy. FDA would proceed against the manufacturer. For both the press and the food men, previously proposed power which allowed the government to establish multiple quality grades on food products was withdrawn. Only the minimum grade concept embodied in the 1930 McNary-Mapes Amendment was retained. As a general concession to quell trade criticism of S. 1944, regarding the "excessive" power of the Secretary of Agriculture the new bill provided for creation of two advisory boards. These boards could act as appeal bodies on regulations proposed by the Secretary.

Copeland called S. 2800 "a sane, sensible, workable bill," but many segments of the concerned industries still had doubts. Most of the objections launched in the fall of 1933 continued unabated. Manufacturers gloomily predicted economic ruin if the new bill passed. It was still anti-NRA. It would still establish a dictatorship over the food and drug trade, and it was still a "plot" of the AMA. Finally, insofar as much of the trade was concerned, S. 2800 retained the ominous "Tugwell" tag.[9] Yet the concessions

[8] *Congressional Record*, 73rd Cong., 2nd Sess. (February 19, 1934), 2728.

[9] *Ibid.*, 2729; "Experts Also Spoke," *Survey* 70 (January 1934), 16; "Al Smith on the Tugwell Bill," *PI* 166 (January 4, 1934), 32; "Observations of the Day," *Standard Remedies* 21 (May 1934), 3; "Tugwell's Trojan Horse," *Plain Talk* 3 (March 1934), 3.

made were real. The trade knew this, and time did bring
some changes in the position of the affected industries.
Many groups in the business community were more satisfied with S. 2800. Concessions to periodical publishers
had not eliminated all their objections but had at least
taken into account the most outstanding ones. A significant
number would continue to oppose Copeland's bill, but in a
much more subdued fashion. The food men, who originally
constituted the least adamant part of the opposition to
revision, were much mollified by the changes embodied in
the new bill. They still had some grievances, but mainly
over formula disclosure requirements which remained on
their products. Charles Dunn and other spokesmen for the
industry strongly protested this point at the February
hearings. Their argument was not militant, however, and
left the door open to compromise.[10] They were soon to win
additional concessions.

The patent-medicine industry provided the spearhead
and heart of the 1934 opposition to efforts at revision of the
1906 law. Even here there was a noticeable change in the
nature of the rhetoric. Concessions made by FDA
prompted, indeed demanded for purposes of public image,
proprietary manufacturers to admit that S. 2800 was a better bill than S. 1944. The blanket assault on the philosophy
of the "Tugwell" bill would be less effective now. To talk
simply of "bureaucracy" was no longer enough. "In other
words," the *New Republic* mused, "they are saying if we
must have a law, at least do not let anybody enforce it."[11]
It was inadequate also to growl indignantly about the "right
of self-medication." "The phrase, itself, sounds synthetic,"
commented *Printers' Ink*, "the man who buys a box of pills

[10] *Congressional Record*, 73rd Cong., 2nd Sess. (February 19,
1934), 2728-29; *Food, Drugs, and Cosmetics, Hearings before the
Committee on Commerce*, U.S. Senate, 73rd Cong., 2nd Sess. on S.
2800 held February 27–March 3, 1934, testimony by Dunn, 56-57.
Hereafter cited as *Senate Hearings on S. 2800* (1934); L. V. Burton
to Paul B. Dunbar, November 26, 1934, Correspondence, Box 429.
[11] "The Week," *New Republic* 78 (February 14, 1934), 3.

seldom feels that he is . . . a crusader in the cause of human liberty."[12]

The generalized assault on a new bill had been effective, and it would never be totally dropped, but by 1934 it became incumbent upon opponents to be much more specific in their objections. The old argument that no revision of the law was necessary was replaced in emphasis by the assertion that the real trouble in FDA was a lack of appropriations. If appropriations were increased, such weaknesses in the law as remained could be corrected by amendments. The trade, it was argued, was certainly receptive to any "needed" amendments.[13] Indeed, the trade actively sought one change, a shift of the law-enforcing agency from the Agriculture Department to the Commerce Department. USDA was set up to aid the farmer, not business.[14] The move would relieve the trade, also, of Rexford Tugwell and his coterie of New Deal liberals currently ensconced in the Agriculture Department.

The advisability of the amendatory approach was borne out supposedly by the many "vicious" provisions which remained in S. 2800. Contrary to the opinion of the Solicitor's office in USDA, the trade charged that the current bill would negate the accused's right of appeal to the courts. There were still too many ambiguous definitions in the measure, despite FDA's contention that the major definitions attacked were lifted almost verbatim from the 1923 Supreme Court decision in *U. S. v. 95 Barrels of Vinegar*.[15] Nor were the proprietary men happy over the current bill's demands regarding therapeutic claims in advertising. These were unrealistic and far too restrictive and simply meant it would no longer be feasible to advertise.[16] Ad

[12] "Common Sense on the Tugwell Bill," *PI* 165 (November 23, 1933), 88.
[13] "Observations of the Day," *Standard Remedies* 21 (June 1934), 2; "Life in the Old Law Yet," *Food Industries* 6 (July 1934), 293.
[14] *Drug Trade News*, March 19, 1934, FDA Scrapbooks, Vol. 3.
[15] "Food Administration Stands Pat on Tugwell Bill," *PI* 165 (December 7, 1933), 6ff.
[16] "Food, Drugs, and Poison," *Current History* 40 (April 1934), 38.

man William Groom offered a specimen of a permissible advertisement to a special meeting of the Proprietary Association:

> We think our medicine is good. There may be other better brands, but at least ours is as good as the average. —Thousands of physicians have prescribed it . . . for certain disorders, but we dare not tell you what the disorders are.—If your doctor should prescribe it for a headache . . . you will still have the headache long after you think it is gone.—In spite of all this, we must urge you to buy our medicine anyhow, as we need the money to push our sales of impure food for the purpose of poisoning your children.[17]

Concessions which Copeland had granted on formula disclosure were not sufficient either. Proprietary manufacturers feared that the full disclosure label provisions would soon be revived, and they remained unhappy with the clause which insisted that certain specified ingredients must be indicated on labels. At the February hearings they talked in trembling tones of the piracy of valuable formulas. The curious thing, commented the *Christian Century*, was that, "on questioning . . . it appeared that they all knew each other's formulae anyway. Only the consumer was in the dark." "What an indictment they draw against their own business," that journal continued. "They say plainly that they cannot tell the consumer the truth about their products and remain in business."[18] The comment was unfair as a blanket indictment but not without some substance.

The old albatross of AMA conspiracy, hung round the neck of revision efforts, was transmuted also into more specific form. The inclusion of "devices" under the definition of drugs was a case in point. The source of that addition was clear. "It is well known," Dinshah Ghadialli, Presi-

---

[17] "Tugwell Bill is Assailed," *PI* 165 (October 19, 1933), 88.
[18] "Self-Portraiture at Washington," *Christian Century* 51 (March 21, 1934), 390-91.

dent of the Spectro-Chrome Institute, stated at the February hearings, "that the medical trust have always opposed every other system or method of healing."[19] The establishment of the two review boards came in for its share of trade criticism partly on the same basis. Creation of the boards was important, opponents of S. 2800 admitted, but the specification that members must have no financial interest in the affected industries was not good. They wanted members of their own community to be represented in those two bodies. As things stood, the opposition asserted, control would be placed at "535 North Dearborn Avenue, Chicago"—home offices of the medical profession.[20] The first grievance probably had substance, even if exaggerated, but the second charge was simply another attack on the ever useful AMA bogeyman.

So opponents of food law revision became more specific and concrete in their criticisms, but no less effective. Doubtless many segments of all the affected industries were willing to accept a new law and were seeking presently only what they deemed to be legitimate business rights. Opposition in the trade was never a concerted "plot" to wreck revision. Yet the irreconcilables were there, particularly within the proprietary medicine field. That camp was smaller by the end of the year, but they carried 1934 with ease.

The basic obstacle for proponents of revision was public apathy, the traditional nemesis of reformers. "All is quiet on the public front," quipped Robert Swain in *Drug Topics*.[21] The killing disinterest was bad enough, but even this was not the whole problem. Among those who were following the fight for S. 2800 some felt the measure was too harsh. Dr. Walter Alvarez of the Mayo Clinic believed there was still too much power lodged with the Secretary. Ironically his reservations were based on memories of the

[19] *Senate Hearings on S. 2800* (1934), testimony by Ghadialli, 59.
[20] *Ibid.*, testimony by Clinton Robb, 132; "No Financial Interest," *D&C Industry* 34 (January 1934), 15.
[21] December 10, 1934, FDA Scrapbooks, Vol. 4.

Harvey Wiley regime. Wiley meant well, Alvarez wrote Swann Harding, but the Minnesota physician had qualms about an office with so much potential power at some point falling to another "hide-bound crusader."[22]

More important was the growing body of revision proponents who felt the current bill did not go far enough. The *New Republic* grumbled editorially that if there were one more revision like the last one there would be nothing left to concede.[23] The People's Lobby, a militant consumer group featuring among its leadership such liberal spirits as John Dewey, Stuart Chase, and Oswald Villard, called the bill "emasculated" and unfit for support.[24] Consumers' Research was equally irate, and pronounced the whole proceedings "a delightful puppet show with the quack medicine men and their colleagues manipulating the strings."[25] The current bill was no longer worth backing and, CR asserted, "it has now become a matter of urgent public concern . . . to prevent the passage."[26]

Schlink-Kallet and company pounded that message home in the CR *General Bulletin,* at the February public hearings, in circularized propaganda to the public, even telegrams to federal officials.[27] The scope and impact of the attack was bad news for the prospects of S. 2800. Walter Campbell saw this as did David Cavers.[28] Ruth Lamb with characteristic color put it bluntly to confidant Paul Anderson, "Have you seen this tripe that Schlink has been putting

[22] Alvarez to Harding, February 17, 1934, Papers of T. Swann Harding, Library of Congress, LC III-23-P, 3, Washington, D.C.
[23] "The Week," *New Republic* 78 (March 28, 1934), 170.
[24] Benjamin March to Roosevelt, March 3, 1934, Correspondence, Box 431.
[25] "No New Deal on Food, Drugs, and Cosmetics," *General Bulletin of Consumers' Research* 4 (October 1934), 2.
[26] Ole Salthe to Campbell, April 9, 1934, Correspondence, Box 431.
[27] *Senate Hearings on S. 2800,* testimony by Arthur Kallet, 276-77; Kallet to Roosevelt, March 1, 1934, Correspondence, Box 431; F. J. Schlink to Bailey, March 14, 1934, Bailey Papers; "A Food, Drug, and Cosmetic Bill Note," *General Bulletin of Consumers' Research* 3 (April 1934), 6.
[28] Campbell to Cavers, April 14, 1934, Correspondence, Box 431.

out?" she wrote in April. "If the bill fails to go through we can certainly thank these 'liberals' in large measure."[29]

Campbell had some reservations about the bill too, but Campbell was also a realist.[30] When S. 2800 was reported out of committee in March the drug chief was "delighted" that concessions to the trade were as few as they were.[31] He understood, perhaps better than anyone else, the power and force of the opposition. Well he should understand it and well he should be satisfied with the limited modifications embodied in S. 2800. Officially, at least, FDA had been able to do virtually nothing to assist Copeland in his travail. The agency labored still under the posture of caution forced on it by revision opponents in the late fall of 1933.

The Deficiency Appropriations Act of 1919 hung over the head of FDA like the sword of Damocles. Campbell was rightfully defensive. Apparently in the wake of trade charges regarding the impropriety of agency activities even Franklin Roosevelt had directed the regulatory body to remain in the background insofar as a new bill was concerned.[32] Such direct promotional activities as speaking engagements by agency personnel remained in suspension. Indeed, Campbell felt compelled to completely disavow any association with the private efforts of consumers seeking passage of S. 2800. One such effort was a chain letter asking recipients to organize petition campaigns to Congressmen in support of a new food bill. Unfortunately for Campbell, the letter further suggested that lists of those solicited be sent to FDA with the request that agency literature on the shortcomings of the 1906 statute be shipped to each addressee. The drug unit chief ruled that such requests could not be honored. To participate in this "scheme" would violate the spirit of the Deficiency Appropriations Act.[33]

[29] Lamb to Anderson, April 14, 1934, *ibid.*, Box 438.
[30] Campbell to Tugwell, February 21, 1934, *ibid.*, Box 431.
[31] Campbell to Tugwell, March 17, 1934, *ibid.*
[32] Lamb to Freda Kirchwey, February 7, 1934, *ibid.*, Box 438.
[33] Campbell to Dunbar, January 29, 1934, *ibid.*, Box 430.

Paul Dunbar regretfully, but dutifully, even rejected the pleas from the women's organizations for agency literature when it appeared that the material was to be used in conjunction with resolutions to Congress in support of a new law.[34] The pressure on FDA came in many forms. Texas Congressmen condemned the regulatory body for placing Crazy Crystals in the "Chamber of Horrors." Representative Blanton boasted that "we" forced the withdrawal of the Texas manufactured mineral water from the exhibit. Campbell denied that the agency yielded to political pressure, but the denial was not altogether convincing.[35] Crazy Crystals had been removed. Walter Campbell knew all about the power of the opposition.

FDA continued to use the "Horrors" exhibit with good effect. Its value was simply less now because the display could not be loaned except upon unsolicited request and only then when it would not be used as part of efforts to influence Congress. Yet where it was used the results were good. Certainly House members were properly impressed when Representative William Sirovich took portions to the floor of the chamber for his colleagues to see firsthand.[36] The impressive effect of the display was undoubtedly a chief motivation in Ruth Lamb's decision to begin a book based upon the exhibition. The book would not be published until 1936 but word of the pregnancy was around long before.[37] The trade was unhappy about the prospects of publication just as they were unhappy about the current exhibit. But the fruition of Miss Lamb's literary efforts was two years hence. In the year 1934 the opposition had too many reasons to rejoice to long be concerned about Miss Lamb. One was the restricted posture of FDA. One was the alienation of militant consumer groups from S. 2800.

[34] Dunbar to Mrs. T. R. Fisher, February 9, 1934, *ibid.*
[35] Kallet to Campbell, May 21, 1934, and Campbell to Kallet, May 25, 1934, *ibid.*, Box labeled Item 17.
[36] *Washington Post*, March 7, 1934, FDA Scrapbooks, Vol. 3.
[37] Louis Engel to Lamb, December 20, 1934; Lamb to Karl Kahn, October 9, 1934, Correspondence, Box 438.

Yet another was the complicated morass of opinions, actions, and inactions within the structure of the New Deal. At the center of the maelstrom was Royal Copeland. Representatives of Consumers' Research had first brought to light the Senator's dealings with Fleischmann's Yeast, Eno Salts, and Phillips' Milk of Magnesia at the December public hearings on S. 1944. This information became a rallying cry for opposition to the New Yorker's sponsorship of drug reform in 1934. Copeland apparently saw no contradiction between his Congressional role and accepting broadcasting fees from the concerns mentioned.[38] In truth these dealings involved no impropriety, and charges to the contrary were specious. The commercial relationship was, however, illadvised and unfortunate. It was harmful to the Senator's reputation and, therefore, it was harmful to the prospects of passing a new food bill.

Kallet and Schlink, who looked with a jaundiced eye on the compromised contents of S. 2800 anyway, dismissed Copeland as a tool of vested interest.[39] They in turn encouraged the following of CR to do the same. Long-standing friends of FDA were also disturbed. The *Nation's* James Rorty wrote to question Ruth Lamb: "Is Copeland working for his bill or is he working for Fleischmann? . . . I intend," he warned, "to tell the facts about Copeland's ambiguous relation to the whole business in the last chapter of my book."[40] Paul Anderson was equally unhappy. He flatly charged in the *Nation* that Copeland had stacked the February hearings in the favor of trade lobbyists.[41] Even Norman Thomas scored the New Yorker in his 1934 book, *Human Exploitation.*[42]

[38] Arthur M. Schlesinger, Jr., *The Coming of the New Deal* (Boston, 1959), 358.

[39] F. J. Schlink, "The Acid Test of the New Deal Liberals," *Common Sense*, September 1934, Correspondence, Box 438.

[40] Rorty to Lamb, April 6, 1934, *ibid.*

[41] Anderson, "Washington Side Show," *Nation* 138 (March 21, 1934), 331.

[42] Bryant Putney to Lamb, December 31, 1934, Correspondence, Box 438; Thomas, 322.

Walter Campbell had confidence in Copeland, but the Senator's "business" dealings were more than a bit embarrassing. FDA information officer Ruth Lamb defended him constantly. "I have no apologies whatever to offer for Senator Copeland," she wrote sharply to Varian Fry of the periodical *Scholastic.* Yet, more defensively, she added, "Whether we approve of his advertising or not we can do nothing about it. After all we can use only such Congressional material as the voters send us."[43] There was also more than a hint in her correspondence that Copeland's continued sponsorship was not an unmixed blessing. Louis Engel, editor of *Advertising and Selling,* wrote to ask if the Senator would handle the Administration's food-drug reform measure in 1935. Miss Lamb replied that she presumed so, basing this conclusion significantly on strong feelings in the upper house about Senatorial courtesy. The thing to do, she concluded, again significantly, "is to get the Progressives so interested that no matter who sponsors the bill they will make an effort."[44]

The credibility of the charges made against Copeland by his most consumer-oriented critics was enhanced by his seemingly easy willingness to compromise with opponents. This was a long-standing personality trait. Perhaps the inclination was the mark of a successful legislator, but to the liberal spirit it provoked utter exasperation. At the December hearings on S. 1944 the opening statement was delivered by Henry Wallace rather than Rexford Tugwell, who had sparked the food-drug reform drive. Some trade commentators interpreted this as a move by Tugwell to disassociate his name from the bill and thereby make it more palatable to the trade.[45] It was just as likely, however, that the absence of the Assistant Secretary was the first indication of Tugwell's growing dissatisfaction with Copeland. Tugwell's liberal spirit was one of those which burned

[43] Lamb to Fry, October 1, 1934, Correspondence, Box 438.
[44] Lamb to Engel, December 7, 1934, *ibid.*
[45] "Down, Fido!" *Food Industries* 6 (January 1934), 3.

with indignation at the Senator's compromising temperament.

This division complicated further the travail of passing a new law. Tugwell openly labeled S. 2800 "very disappointing" and called for a strong public campaign to revitalize the bill.[46] By April 1934 he had stated repeatedly that the measure had been so watered down as to cause him to lose all interest in it.[47] The *Christian Century* claimed Tugwell's March trip to Puerto Rico was taken in disgust over the fate of the food measure.[48] In truth, Tugwell was very upset over compromises in Copeland's measure, to the point that in February he had taken his grievances direct to Roosevelt. He charged that Copeland had allowed the enemies of a strong law to rewrite the bill to suit themselves and announced his intention to withdraw personal support from S. 2800.[49] The revision effort was no longer an honest one. Yet, for all this rhetoric, Tugwell never totally lost his interest in a new bill, even in its weakened form, but he remained unhappy about the course of events.

His state of mind proved an unfortunate circumstance, for Tugwell was the spur to Roosevelt's interest in the bill. Copeland's voice alone was not adequate. The President's relationship with revision efforts was nothing short of complete enigma. Again and again observers of the struggle reported that only the intervention of FDR would bring passage. That leadership never came, and when S. 2800 died on the Senate calendar much of the responsibility could be placed with Roosevelt. The President's failure to act with decision had many complexities. For one thing, he too was subject to heavy trade pressure. There were also those in his party, such as Representative Wesley Disney of Oklahoma, who urged him to defer the measure

[46] "The Week," *Nation* 77 (January 17, 1934), 264-65.
[47] *OP&D Reporter*, April 16, 1934, FDA Scrapbooks, Vol. 3.
[48] "Further Adventures of the Pure Food Bill," *Christian Century* 51 (March 28, 1934), 413.
[49] Tugwell to Roosevelt, February 21, 1934, OF 375, Roosevelt Papers.

until the next session lest its passage hurt Democratic prospects in the fall elections.[50]

There was also a brewing jurisdictional tangle between FDA and the Federal Trade Commission destined to be one of the most important aspects of the five-year battle for a new law. The first public indication of this squabble showed up when Judge Ewin Davis, chairman of the FTC and a past Representative from Tennessee, appeared at the February hearing to protest the advertising clauses of the Copeland bill. Davis claimed that advertising control came under the jurisdiction of the Federal Trade Commission and was being adequately handled by his agency. Copeland argued that S. 2800 proposed no jurisdictional overlap, since the Raladam decision of 1931 limited FTC's advertising controls to unfair competition. The new FDA power would apply to consumer protection against false advertising. Davis was unmoved. He dismissed the Raladam case as a "rare exception" and insisted that false advertising almost invariably involved unfair competitive practices.[51] FTC carried this contention to Roosevelt even before the hearings. Tugwell in turn took up the defense of FDA. Presumably the decision on who got controls over advertising was left to the Chief Executive.[52]

Theoretically this decision would take time and so help to explain the President's immediate failure to press for passage of the Copeland measure. Unfortunately, the *real* answer to the failure of decisive executive leadership, which would plague food bill revision efforts all along, does not rest in such concrete considerations. Indeed, FDR remained secretive and equivocal on the jurisdictional dispute. It was settled ultimately in the House with no direction from the President. The most apparent key to Roosevelt's curious posture is that he was never enthusiastically committed to the passage of a new food bill. This legisla-

[50] *Ibid.*, Wesley Disney to Roosevelt, January 26, 1934, *ibid.*

[51] *Senate Hearings on S. 2800* (1934), testimony by Davis, 233-34.

[52] Tugwell to Roosevelt, January 26, 1934; Roosevelt to Tugwell, January 22, 1934, OF 375, Roosevelt Papers.

tion remained outside a rather narrowly defined conception of national reform needs, though the President did become slightly more responsive as the New Deal veered to the left after 1935.

The history of the 1938 law also supports historian Richard Hofstadter's criticism of Roosevelt's free and easy view on the responsibilities of power. As Hofstadter has quite properly pointed out, FDR "suffers by comparison with men of the spiritual gravity of Lincoln and Wilson." The President's "grasp of his responsibilities was neither intellectually nor spiritually profound."[53] Thus, his lukewarm desire to see a new food bill pass was dampened further by the fact that Copeland sponsored the measure. The New York Senator was a New Dealer by party affiliation rather than philosophical inclination. By 1934 philosophy began to win out, and he joined the opposition on several administration measures including the Independent Offices bill. He was increasingly on the outs with FDR, a fact which discouraged the President from supporting Copeland's pet measure—the food bill. Friction with the administration reached such a peak by the fall of 1934 that FDR's political right arm, James Farley, publicly refused to support Copeland for reelection.[54]

Nor did Roosevelt offer election aid, and Copeland felt this "Presidential snub" strongly. Drew Pearson reported in his "Washington Merry-Go-Round" column that the Senator was so upset he had decided not to consult with the administration in the preparation of yet another revision of the food and drug bill.[55] One may easily surmise that this report was hardly a matter of great concern to the President. In public and in private he displayed little interest in the

[53] Hofstadter, "The Roosevelt Reputation," in Morton Keller, ed., *The New Deal, What Was It?* (New York, 1963), 21. For a wider treatment of the view see Hofstadter, "Franklin D. Roosevelt: The Patrician As Opportunist," in *The American Political Tradition* (New York, 1948).

[54] Oswald Villard, "Issues and Men, Let Us Abate Senator Copeland," *Nation* 138 (April 18, 1934), 433.

[55] *New York Daily Mirror*, December 15, 1934, FDA Scrapbooks, Vol. 4.

fate of the measure throughout the year. Even the rift between Copeland and Tugwell at the opening of 1934 was not taken very seriously. FDR called both in during February, directing a private secretary to arrange an appointment for two men to fight it out and he would "sit in and hold the sponge."[56] There is no indication he did more.

The President was questioned about the state of food-drug law revision at a press conference in the same month. There were "three or four" bills in Congress, he replied. "What will come out from them I do not know." He then declined to state a preference for any of them.[57] The only publicized aid which Roosevelt gave the Copeland bill throughout the 1934 Congressional sessions was an April memo to Senate majority leader Joseph Robinson. This was done in such an unobtrusive way that few people were clear on what the note actually said. Ruth Lamb for a time was not even sure that it existed.[58] All the message did contain was a transmittal of Copeland's request for time, time to bring up S. 2800 for a vote. FDR's only comment was that he told the New Yorker he had no objection if Copeland "could get his bill through without holding up the session."[59]

Questioned in May by the press about the Robinson note and S. 2800, Roosevelt stated simply that he did not know what would happen to the Copeland bill.[60] This noncommittal attitude by the President was taken for what it was worth. The press concluded that the President was not now pushing for passage.[61] *Printers' Ink* summed up the consensus: "President Roosevelt seems to believe that the

[56] Marguerite LeHand to Marvin McIntyre, February 28, 1934, OF 375, Roosevelt Papers.

[57] *Roosevelt Press Conferences*, February 2, 1934, Vol. 3, 129-30, Roosevelt Papers.

[58] Lamb to Engel, June 1, 1934, Correspondence, Box 438.

[59] Roosevelt to Robinson, April 19, 1934, OF 375, Roosevelt Papers.

[60] *Food Field Reporter*, June 6, 1934, FDA Scrapbooks, Vol. 4.

[61] *Ibid.*; *Christian Science Monitor*, March 5, 1934, *ibid.*, Vol. 3.

country can struggle along under the present Food and Drug Act for a few more months." The consensus hardly inspired decisive action in Congress.[62]

Such pressure and support for passage as was felt in the national legislative halls in 1934 came largely from sources other than the official family of the New Deal. The vast body of the American public remained apathetic, but the food law revision camp did gain in strength during the year. There were the additions made by men like Stuart Chase who decided to make the drug issue personal and dramatize it in his lecture tours. Chase considered coming to join the fight in Washington but concluded his name would have the same adverse effect as that of Tugwell.[63] There was a growing number of professional organizations such as the American Society of Biological Chemists who put themselves on public record as favoring revision.[64] The American Medical Association took a more positive stand in both *Hygeia* and *JAMA*. A March editorial in the latter challenged members to support S. 2800 by direct contact with legislators and through personal contact with the public.[65]

The main credit for effective support of the Copeland bill went to the national women's organizations. Their mobilization behind revision began in the fall of 1933, and it grew in force throughout 1934. By spring the number of national bodies rose to ten when the General Federation of Women's Clubs wired FDA of its backing.[66] According to *Printers' Ink* these "militant women . . . almost ran away with the show" at the February hearings.[67] They were not happy with many of the compromises in S. 2800, particu-

[62] "Copeland Bill Hits Snag," *PI*, 167 (May 24, 1934), 65.
[63] Chase to Tugwell, February 10, 1934, Correspondence, Box 438.
[64] "Medical News," *JAMA* 102 (May 26, 1934), 1771.
[65] "The Advertising of Foods, Drugs, and Cosmetics," *Hygeia*, 12 (January–February 1934), 6-7; "The Tugwell-Copeland Pure Food, Drugs, and Cosmetic Bill," *JAMA* 102 (March 3, 1934), 696.
[66] Lamb to Winifred Willson, May 29, 1934, Correspondence, Box 438.
[67] "Consumers in Limelight at Copeland Hearing," *PI* 166 (March 1, 1934), 20.

larly the loss of provisions on quality grading standards. At the hearings they fought hard for restoration. Miss Alice Edwards keynoted the assault by introducing an endorsement of the grading clauses obtained from Eleanor Roosevelt.[68]

The women continued to demand a stronger bill after the hearings. What distinguished them from militant bodies, such as Consumers' Research, was that failure to get all they wanted had little effect on their zeal. Even a weakened new law would be better than none, and the women felt there were many good features remaining. The advertising control features were cases in point. The only qualms one might have about them, wrote Catherine Hackett, was that existence would lose a little of its adventure. "I can hardly imagine life without the daily thrills of fear of pyorrhea . . . or the thrilling realization that I can retain Henry's ardent affection by using turtle-oil to stave off wrinkles." She was prepared, however, to live with that loss.[69]

The battle lines of the women took many forms. Organizational magazines kept readers up-to-date on the status of the bill and urged them to action.[70] The women circularized consumers, often using purchased reprints of strong editorials by James Rorty and other spokesmen for reform.[71] Most of all they made their sentiments known to their elected representatives through organized write-in campaigns. Often their correspondence was about the only endorsement of the food-drug bill which Congressmen received.[72] FDA was delighted with the aid and even more so with the effect. The drug unit's Charles Crawford wrote with satisfaction to David Cavers that the women "have

[68] *Senate Hearings on S. 2800* (1934), testimony by Edwards, 85.
[69] Hackett, "A Housewife Praises," *Forum* 91 (February 1934), 99.
[70] "Go To It, Home Economists," *Jnl. of Home Economics* 26 (October 1934), 520; Harriet Howe, "Food and Drug Legislation and Home Economics Clubs," *ibid.*, November 1934, 559-60.
[71] Lamb to Rorty, February 24, 1934, Correspondence, Box 438.
[72] See as example the correspondence of Senator Bailey, Bailey Papers.

become so active that some of the trade interests are really becoming alarmed."[73]

The trade was concerned. They were concerned about the growing voice of revision demands in general—concerned about how continued militant opposition to these demands might affect their retail markets.[74] Manufacturers were concerned also by their own increasing suspicion that some food bill was going to pass. If they were to continue to fight the Copeland measure that opposition must take a more positive form. In lieu of S. 2800 the trade must submit their own reform measures. One answer was a drive for self-regulation. This move was spearheaded by segments of the drug industry which had been most militant in opposition to a new law, the United Medicine Manufacturers Association and the Proprietary Association. Their reasoning was clear.

Drug man Lee Bristol told the former organization in October that there was an absolute necessity for setting up some system of self-control over advertising claims or face "drastic government censorship."[75] Two months before, Frank Blair, president of the Proprietary Association, went a step further. He announced that the organization had established an Advisory Committee on Advertising which would pass on the advertising of members. The motivation was the concern about future legislative action.[76] A complementary strategy, though begun earlier, was for the concerned industries to submit their own versions of statutory revision.

This plan seemed a good one. For irreconcilables the approach would hopefully confuse the issue and postpone passage indefinitely. For the less adamant there was the hope that one of the weaker trade bills would pass, or, if not, that these bills might serve as a lever to win further concessions from the administration.[77]

[73] Crawford to Cavers, March 9, 1934, Correspondence, Box 431.
[74] R. M. Allen to Campbell, March 1, 1934, *ibid.*, Box 429.
[75] *Drug Trade News*, October 29, 1934, FDA Scrapbooks, Vol. 4.
[76] "Proprietaries Adopt Code," *PI* 168 (August 16, 1934), 24.
[77] "Too Many Food Bills," *PI* 169 (November 19, 1934), 104.

Of the substitute measures two were especially threatening to the hopes of FDA. One was a series of amendments known as the Beal bill. It had been introduced in the House by Representative Loring Black of New York and claimed the support of important segments of the proprietary medicine industry. Typically attractive provisions included enforcement by cease and desist procedure rather than criminal prosecution, and the wide latitude given to therapeutic claims permissible under the guise of "opinion" rather than "fact." Paul Anderson declared in disgust that the measure ought to be titled, "a bill for the protection of fakes and the promotion of mass poisoning."[78]

A more threatening substitute was the so-called McCarran-Jenckes bill, which had been drawn by Charles Dunn, counsel for the Associated Grocery Manufacturers of America. The measure was chiefly backed by that organization, the American Pharmaceutical Manufacturers, and the National Drug Trade Conference.[79] Dunn's proposal was a greater threat than the Beal bill because it was a better bill, in many ways not unlike Copeland's offering. The "ringer" was that enforcement was highly complicated and provisions existed for an almost endless round of appeals. Copeland charged that no less than 25,000 people would be required just to administer such a law.[80] Ruth Lamb called it "Dunn's celebrated opus for the relief of indigent and unemployed lawyers."[81] Whatever the weaknesses, both bills had wide attraction.

Trade ballyhoo put that appeal before a large audience. Congressmen like North Carolina's Senator Bailey, who were troubled by the "drastic" provisions of the Copeland measure, found such substitutes particularly attractive. Bailey was converted to the Senate version of the Beal bill

[78] *St. Louis Post-Dispatch*, January 22, 1934, FDA Scrapbooks, Vol. 3.

[79] "The Drug Act Situation," *D&C Industry* 34 (February 1934), 116.

[80] Charles Dunn, *Federal Food, Drug, and Cosmetic Act*, 91.

[81] Lamb to Rorty, February 15, 1934, Correspondence, Box 438.

by March 1934.[82] Copeland became concerned enough about the popularity of the Dunn bill to participate in debates with its author over nationwide radio.[83] Yet if the confusion of bills and views was a menace to the hopes of FDA, by the late fall of 1934, some segments of the trade began to feel they were being hoist with their own petard. *Printers' Ink* viewed the hoard of bills in Congress with alarm. Some measure was going to pass, it warned. Manufacturers had better get together on what they wanted, or they might be very sorry in the future.[84]

The industry versions were not the only ones being presented in the Congressional halls. There was the Boland bill, prepared by Consumers' Research, and it was filled with dynamite. Not only did the bill demand full label disclosure of formulas and rigidly regulated advertising, but it called for an industrial licensing system that would make manufacturing operations totally dependent on the good will of the government.[85] There were also peripheral proposals such as the Huddleston bill. This was a branding measure which would lodge power in the Bureau of Standards to establish a full-blown quality grading system for all goods in interstate commerce. If the bill should pass, one trade journal gloomily prophesied, it would kill trade marks, trade names, and much of the return from advertising.[86]

The appearance of such radical measures, though they were not likely to pass at present, provided fearsome food for thought about the future. The growing popularity of men like F. J. Schlink was also thought-provoking. *Printers' Ink* was shocked at the response he got at an appearance before the American Academy of Political and Social Science, and argued that something had to be done. If such

[82] Musterole Company to Bailey, February 20, 1934; Bailey to Dizor Drug Store, February 28, 1934, Bailey Papers.

[83] *Drug Trade News*, March 19, 1934, FDA Scrapbooks, Vol. 3.

[84] "Too Many Food Bills," *PI* 169 (November 19, 1934), 104-105.

[85] *Drug Trade News*, March 5, 1934, FDA Scrapbooks, Vol. 3.

[86] "And Now the Huddleston Bill," *PI* 166 (January 4, 1934), 83; "The Huddleston Bill," *ibid.*, January 11, 1934, 49-50.

appeal among the professorial class continued, "in thousands of classrooms in colleges all over the country young men and women are going to be educated to believe that honesty and advertising can't mix."[87] One way out, the journal concluded by the fall of 1934, was to get a new food law on the books before things got worse.[88]

Many segments of the affected industries were reaching the same conclusion. *Food Industries* expressed the sentiment in June: "As a matter of practical politics we are veering to the opinion that the food industries would be better off if the revision . . . were enacted this summer than to take a chance on what may happen next year."[89] Even Charles Dunn asked his Associated Grocery Manufacturers of America to withdraw their past objections to the advertising provisions of Copeland's bill.[90] The publishing industry itself had been much more content with S. 2800 than S. 1944. By the fall, much of that trade was actively pushing for the early passage of a new law. One factor was a matter of economics. Sizable numbers of advertisers were refusing to renew contracts until the business of a new law was settled.[91]

A most significant defection from the opposition camp in 1934 came within drug circles. There were, indeed, certain drug interests that had never truly opposed a stronger law per se. These groups represented the more professional side of pharmacy as opposed to commercially oriented organizations such as the National Association of Retail Druggists or the National Wholesale Druggists Association. One organization in the former category was the American Association of Colleges of Pharmacy. The Association had not taken an opposing stand toward the Tugwell bill even though some of its members did express the view that certain features of S. 1944 were too drastic. One of its particular concerns was the wide discretionary power pro-

[87] C. B. Larrabee, "Mr. Schlink," *ibid.*, 10, 13.
[88] "The New Food Bill," *ibid.* 169 (December 20, 1934), 88.
[89] "New Food Law." *Food Industries* 6 (June 1934), 2.
[90] "Advertising Law for Food," *PI* 169 (November 29, 1934), 65.
[91] Lamb to Rorty, February 15, 1934, Correspondence, Box 438.

posed for the Secretary of Agriculture. After such features were modified in Copeland's bill, S. 2800, that group went on record for passage.[92] The *American Journal of Pharmacy*, a publication of the Philadelphia College of Pharmacy and Science, editorially labeled the 1906 law inadequate and cheered the prospects of a new act as early as September 1933.[93] The American Pharmaceutical Association had qualms over some parts of the Tugwell bill but neither was that organization opposed to stronger drug controls.[94]

In the immediate months of struggle over the Tugwell bill such professional bodies generally sought some type of unity with their more commercial brothers. The former group, as members of the National Drug Trade Conference, duly subscribed to Dr. James Beal's criticism of the original proposal, S. 1944. The purpose of the National Drug Trade Conference was to bring about concerted action by drug organizations on matters of general interest. It seems that the more professional groups were initially reluctant to split the Conference, even if their views differed from those of other members.[95] Moreover, the professional bodies were not immune to that natural human conservatism which rebels against the prospects of sudden and uncertain change. Aside from specific objections to matters like the powers given by the Tugwell bill to the Secretary of Agriculture, these groups hoped that revision of the 1906 statute could be carried out by amendment.[96]

The December 1933 issue of the *Journal of the American Pharmaceutical Association* concluded that the National

[92] "American Association of Colleges of Pharmacy and Proposed Food and Drug Legislation," *American Journal of Pharmaceutical Education* 2 (January 1938), 88-89. Hereafter cited *AJPhE*.

[93] Ivor Griffith, "The New Deal for Food and Drug Control," *AJP* 105 (September 1933), 429-32.

[94] "Resolution Number Four," *Journal of the American Pharmaceutical Association*, 22 (September 1933), 880. Hereafter cited *AJPhA*.

[95] See for example, "Food, Drug, and Cosmetic Legislation," *ibid.*, December 1933, 1210-13.

[96] *Ibid.*

Drug Trade Conference had done "a splendid work for pharmacy." "Every pharmaceutical interest should support the Conference."[97] By the fall of 1934, however, the professional bodies, including the APhA, were not so enthusiastic. The fact was that these groups had, for a time, simply acquiesced in the view of the drug manufacturers and their allies in the Conference. The result was not a happy one. Robert Fischelis, President of the APhA, made the point in September.

The Association had traditionally fought for strong controls over food and drugs, Fischelis stated, yet, "last year due to lack of knowledge of the facts or misinformation, the profession of pharmacy was classified as an opponent of new legislation." The National Drug Trade Conference was a valuable organization, but "when interests represented in the Conference indulge in the type of propaganda which was used last year to defeat proposed food, drug and cosmetic legislation, we must take steps to make our position clear." The Association must, he concluded, "let it be understood in no uncertain terms that we are not a party to such a program."[98] Fischelis' comments were quickly repeated with approval in the *American Journal of Pharmacy*.[99] From 1934 forward the more professional organizations, led by the American Pharmaceutical Association, National Association of Boards of Pharmacy, and the American Association of Colleges of Pharmacy, took an increasingly positive and vocal stand on a new food law both in and out of the National Drug Trade Conference.

It is possible that the altered posture of these groups would have come simply as a result of changes made in the 1934 versions of the proposed food and drug legislation. Yet there were interest differences between the professional pharmaceutical associations and the more commercially oriented drug groups which probably played a role

[97] *Ibid.*, 1213.
[98] Fischelis, "Our Stand on Federal Drug Legislation," *ibid.*, 23 (September 1934), 862.
[99] Editorial, *AJP* 106 (October 1934), 372-73.

in the change. Surely the association of the professional bodies with the antics of the drug manufacturers was one. Such an association hardly enhanced the prestige of pharmacy as a profession. Moreover, bodies like the American Pharmaceutical Association, did have a sincere and primary concern with public health as well as drug safety. They were also relatively immune to trade pressures.[100]

Finally, the provisions of the various food bills touched the operation of legitimate pharmacy far less than they affected drug manufacturers. The provisions also affected packaged remedies more than prescription drugs. There was no reason why the representatives of professional pharmacy should feel obliged to crusade in the behalf of proprietary medicines. Quite the contrary. Indeed, the American Pharmaceutical Association had opposed the excesses of proprietaries since long before the enactment of the 1906 law. The sale of such material, while profitable, did not enhance the status of pharmacy, a matter of deep concern to the professional organizations. Proprietary sales were not even limited to the drugstore counters.

Patent medicine interests were particularly unhappy about the growing division within the industries to be affected by new food and drug legislation. In proprietary circles the desire to fight on remained much stronger than elsewhere. Yet even in this group a split existed between those who would accept "reasonable" revision and diehards who were out to block any new legislation. That difference in view would continue to exist in 1935. The total obstructionists would still have their days ahead. Final passage of a law was three-and-one-half years away. The main lines of future tactics, however, would be to "water down" Copeland's bill as much as possible before it passed.

In the meantime, and more immediately, the medicine men were not happy about the new year. Rightly or wrongly they were convinced that FDA had cracked down

[100] Editorial, *AJPhE* 1 (January 1937), 98.

on them because S. 2800 had failed to pass. A "pitiless campaign," one trade journal called it. The drug unit was behaving like a "sulking petulant" child, declared another. "In a blind fury they have started to swing the clubs at the heads of those who were mainly responsible for thwarting their plans."[101] Whatever the truth of this perception the continuing anger of FDA was not a pleasing prospect. As for the views of the White House, there were rumors that FDR failed to push S. 2800 because he intended to try for a stronger measure in 1935.[102] In December, New Deal *bête noire* Rexford Tugwell returned from a trip to Europe. Yes, there would very likely be a new bill introduced in the next Congressional session, he told the press, and yes, he hoped it would be stronger than S. 2800.[103] The industries could well view the future with alarm.

[101] *Drug Trade News*, July 9, 1934, FDA Scrapbooks, Vol. 4; "Sulking Drug Control," *D&C Industry* 35 (September 1934), 244.
[102] *Philadelphia Daily News*, May 30, 1934, FDA Scrapbooks, Vol. 4.
[103] *National Consumer News*, December 1, 1934, *ibid.*

IV

"I LIKE TO THINK ABOUT THE STAR CANOPUS"

When quacks with pills political would dupe us
When politics absorbs the live long day
I like to think about the star Canopus
So far, so far away.

"CANOPUS"
BERT LESTON TAYLOR
(1866-1921)

IN January 1935, what *Printers' Ink* called the "annual food, drugs and cosmetic regatta" got underway once more.[1] By the first days of April the food-drug bill was in serious trouble. Royal Copeland was angry and exasperated. On the Senate floor he denounced the "slimy serpents" of the proprietary medicine trade now "wiggling around this capitol."[2] In disgust he threatened to give up his sponsorship of the pending measure.[3] In the evening sky beyond Sirius and 540 light years from earth was the star Canopus. Unfortunately, it was not visible in Washington. Canopus might have been very consoling to Copeland, "so far, so far away."

S. 2800 had died in committee with the close of the 1934 Congressional session. The New York Senator had gone to work almost immediately on a new bill—his bill. Forgoing the demands of Tugwell and even the assistance of the Food and Drug Administration he had introduced his own version of drug reform in January. It was designated Sen-

[1] "Copeland Bill Is Introduced," *PI* 170 (January 10, 1935), 13.
[2] "Pure Foods Can Wait," *New Republic* 82 (May 1, 1935), 328; *Congressional Record*, 74th Cong., 1st Sess. (April 5, 1935), 5137-38.
[3] Charles W. Dunn, *The Federal Food, Drug, and Cosmetic Act,* 426.

ate bill 5. The wind was at his back. In the fall elections of 1934 Copeland had survived the failure to gain FDR's endorsement for reelection. The food and drug trade was impressed. He would "probably be a Senator long after Roosevelt leaves the White House," *Food Industries* recorded with some admiration.[4] Mississippi's Senator Stephenson, a past stumbling block in the handling of Copeland's pet bill, had been defeated in the fall. The New Yorker would succeed him as Chairman of the Senate Commerce Committee, the committee handling S. 5.[5]

Copeland had made concessions in his new bill. Previous provisions for voluntary factory inspection were gone. The list of diseases for which advertising claims were prohibited had been shortened. Labeling demands on proprietary medicines were more lenient. No longer did labels have to bear the designation as palliative rather than cure. Manufacturers could file formulas with the Secretary of Agriculture and thereby escape label disclosure of contents. By court order FDA could be restricted in misbranding seizures to three actions on a single product. The Senator had also resisted consumer pressure to reestablish multiple grading standards for food products.

Even with the concessions, S. 5 was a good bill and certainly a great improvement over the existing 1906 statute. Copeland knew this, and he was happy to find that FDA felt the same. After study of the measure the agency was prepared to accept it as perhaps a realistic compromise.[6] Copeland retained also the backing of his most effective allies, the national women's organizations. They were not totally pleased about his measure. Alice Edwards of the American Home Economics Association made that quite clear at hearings on S. 5 in March. She had a number of serious grievances. "We have noted with grave concern

[4] "New Food Law," *Food Industries* 7 (January 1935), 3.

[5] *Drug Trade News*, October 1, 1934, FDA Scrapbooks, Vol. 4.

[6] Ruth Lamb to Mrs. James Musser, January 28, 1935, Correspondence, Box 569.

the efforts which are being made to weaken the enforcement provisions of the bill," she told the committee.[7]

Chiefly, Miss Edwards was aggrieved about the failure to restore multi-grading quality standards for food products. She was equally unhappy about the loss of formula disclosure provisions. Finally, she did not like the drug variation clause of the bill. If a product was sold under the Pharmacopeia name, she insisted, no variation from that standard should be permissible, as Copeland's draft allowed. Yet Miss Edwards and the eleven national women's organizations for whom she spoke were as realistic as FDA. They too were willing to back S. 5 with only minor modification, though she did warn that the women's groups accepted the measure as a compromise. They were not willing to see further concessions.[8] At the March hearings representatives from each of the eleven organizations trooped before the committee and endorsed passage. Copeland might well have smiled with satisfaction.

The Senator had originally hoped to avoid new hearings. That effort proved abortive. A scheme had been worked out by mutual agreement between Copeland and Charles Dunn. S. 5 was to be revised in several particulars as a concession to Dunn. He in turn was to give up pushing the so-called McCarran bill, and Copeland's measure would be quietly reported out of committee on February 13. Dunn and Copeland had forgotten, apparently, the skill of the Proprietary Association's counsel, James Hoge. Hoge got wind of the plan and by the hour the committee met, members had been so deluged with telegrams of protest from "back home" that they insisted on new hearings.[9] It was a momentary check. On the whole, things continued to go well for the New York Senator. Even the hearings went

[7] *Food, Drugs, and Cosmetics, Hearings before a Subcommittee of the Committee on Commerce*, U.S. Senate, 74th Congress, 1st Session on S. 5 held March 2, 8, and 9, 1935, testimony by Edwards, 213. Reference hereafter cited *Senate Hearings on S. 5* (1935).

[8] *Ibid.*, 214.

[9] "Copeland Bill Sidetracked," *PI* 170 (February 14, 1935), 18; "Bureaucracy Defeated," *ibid.*, 121.

well. FDA's Charles Crawford wrote David Cavers that the department lost nothing by them. Indeed he had been surprised at the "amount of unqualified endorsement of the bill."[10]

The affected industries had their discontented elements. The immediate weeks after the introduction of S. 5 proved that. The National Broadcasting Company felt that such an important measure should not be rushed through too quickly.[11] The Spectro-Chrome company with their own quackish brand of therapeutic theory still opposed the inclusion of devices within the definition of drugs. Their publications blustered away at the AMA which "is not satisfied with its own fields of poison and pus-punching, but must encroach upon scientific fields in which its ossified cranium can not penetrate."[12] Clandestine groups such as the National Advisory Council of Consumers and Producers, successor to the Joint Committee for Sound and Democratic Consumer Legislation, railed against new legislation.[13] At the March hearings the American Newspaper Publishers Association pressed their view that existing laws on advertising were sufficient.[14] The American Bakers' Association still felt the foods industry deserved a separate law.[15]

As usual, the strongest outcry came from within the proprietary medicine industry. The Proprietary Association, through the journal *Standard Remedies,* was "frankly disappointed." The past objections of the Association to S. 1944 and S. 2800 had simply not been taken into consideration. The new measure was still too broad, and it still

[10] Crawford to Cavers, March 16, 1935, Commissioners' File, Box 11.

[11] Edgar Kobak to Franklin D. Roosevelt, January 19, 1935, OF 375, Roosevelt Papers.

[12] *Spectro-Chrome,* February 1935 (pamphlet), FDA Scrapbooks, Vol. 5.

[13] Louis Engel to Ruth Lamb, April 22, 1935, Correspondence, Box 569.

[14] *Senate Hearings on S. 5* (1935), testimony by Elisha Hanson, 143.

[15] "Food Bill Sidetracked," *Food Industries* 7 (March 1935), 152.

threatened the sacred right of self-medication.[16] Clinton Robb of the United Medicine Manufacturers of America felt that in some ways S. 5 was even more unfair than its predecessors. Before the Senate committee he trotted out all the old criticisms, from the excessive power of the Secretary to the devious workings of the AMA, to the harmful effects of such new legislation on a great industry. He was particularly unhappy with the multiple seizure provisions of Copeland's bill and what he considered to be the "shift" of advertising controls from the Federal Trade Commission to FDA.[17] William Jacobs of the equally adamant Institute of Medicine Manufacturers, in his appearance before the Senate Committee, echoed Robb almost phrase for phrase.[18] What these organizations, representing the hard right of the medicine trade, wanted, was to retain court precedents and the preservation of such uniformity as presently existed between federal and state food-drug legislation—a long-standing trade desire. In fact, as Copeland pointed out to his colleagues, H.R. 3972, the medicine makers' favorite bill, "when it gets through it rewrites the whole act."[19] The real attraction was the pitifully weak enforcement provisions of the bill. The definitions of drugs, cosmetics, and advertising were highly restricted. The enforcement and control powers of FDA, especially in such matters as seizure of goods, were equally restricted. Advertising control was left to FTC with that agency's slow-moving cease and desist procedure which the medicine men much preferred to the criminal-prosecution authority of FDA.

H.R. 3972 or the Mead bill was a carry-over from 1934 trade tactics designed to complicate the legislative scene and delay the Copeland measure. The substitute was dangerous, but the New York Senator could still smile with

[16] "Observations of the Day," *Standard Remedies* 21 (January 1935), 13.
[17] *Senate Hearings on S. 5* (1935), testimony by Robb, 87-90.
[18] *Ibid.*, testimony by Jacobs, 51-52.
[19] Dunn, *Federal Food, Drug, and Cosmetic Act*, 464.

satisfaction. If the Mead bill reached the Senate, it was not likely to pass or even get out of committee. The very best its sponsors could hope for was to force some compromise. Backers of this House substitute were acting more from desperation than real hope that their version would become law. Even *Drug and Cosmetic Industry* admitted in February that the chances of enactment were virtually nil. The real thing was that medicine men felt they had been "ganged" on S. 5. "We have seen Senator Copeland," that trade journal lamented, "sell the drug and cosmetic manufacturers down the river . . . with the aid of the large publishing, broadcasting and food interest."[20]

In a sense *Drug and Cosmetic Industry* was correct, which was precisely why Copeland could smile. The affected industries were capitulating, much to the horror of the hard right in the proprietary medicine trade. The journal was also partly correct in its diagnosis of why they had been deserted. The previous dropping of provisions for multiple grading of food products had proved to be a valuable strategy in winning over the advertising and food industry. On the other hand, the anger of the medicine industry was unrealistic. Perhaps they had taken too seriously their own propaganda about the higher values involved in the opposition toward a new law. Trade unity was not philosophical. It was purely a matter of self-interest, and that fact explained the capitulation.

First and foremost, there was the fear of what the future might bring if the battle went on. Hugo Mock, speaking for the Associated Manufacturers of Toilet Articles, was explicit on the point. His group favored S. 5 because some form of regulation was "in the air." A number of bills "less wisely drawn" had been introduced into the state legislatures.[21] The prescription drug manufacturers were not totally happy with Copeland's bill but the proprietary men

[20] "Ganging the Industry," *D&C Industry* 35 (April 1935), 407; "The Senator's Ambition," *ibid.*, 401; "The Industry's Course," *ibid.*, 36 (February 1936), 133.
[21] *Senate Hearings on S. 5* (1935), testimony by Mock, 96.

[80]

would not be able to count on them for opposition aid. For one thing the nature of the business which the prescription producers conducted, as Swann Harding would point out some years later, forced them to be more scrupulous about appearance and to show more interest in scientific control and research. They could not afford to assume the militant posture of the proprietary interests.[22]

A second factor involved was that many prescription manufacturers felt that their business might be increased if more restrictions were placed on the marketing of proprietary balms.[23] Finally, the chief complaint of the American Pharmaceutical Manufacturers Association and the American Drug Manufacturers Association, representing prescription interests, was the restrictions placed on them by the variation clause in S. 5. As Copeland's bill passed the Senate the ethical producers no longer had to spell out specific variations from standards set in official compendiums. They need only state the standards used on their packaging.[24] Therefore, the opposition of these interests was greatly decreased by the spring of 1935.

Nor could the opposition in any sense expect support from the ranks of the more professional pharmacy groups. That fact was painfully apparent in the dramatic contrast between the testimony of W. J. Schieffelin and Robert Fischelis at the 1935 Senate Hearings. Speaking for the commercially oriented National Wholesale Druggists Association, Schieffelin warned that the current bill was still too broad, urged restrictions on FDA's multiple seizure powers, retention of advertising controls with FTC, and other now-familiar demands.[25] Fischelis, representing the American Pharmaceutical Association, not only called for early passage of a new

[22] Harding, "The Battle for a Better Food and Drug Act," *AJP* 118 (October 1946), 343.
[23] "A Divided Industry," *D&C Industry* 33 (October 1933), 327.
[24] "The A.P.M.A. Meeting," *ibid.* 37 (July 1935), 52; *Senate Hearings on S. 5* (1935), testimony by Horace Bigelow, 206-207; Dunn, *Federal Food, Drug, and Cosmetic Act*, 519.
[25] *Senate Hearings on S. 5* (1935), testimony by Schieffelin, 57-58.

law but argued that formula disclosure on labels should be restored to the current bill. Even a statement of active ingredients was not enough. A layman could not medicate himself intelligently or safely, he declared, if he was not cognizant of the constituents of the medicine.[26] The more professional pharmaceutical groups had simply decided during 1934 that their interests were best served by supporting a strong new law.

This addition to the reform coalition was a bitter pill to the opposition but it was also only one of several unhappy changes within the affected industries. The advertising community had been pleased with the omission of multi-grades in food products, and they had been pleased with prosecution exceptions for unknowingly accepting false advertising. On the whole they had been pleased enough with the latter so that they preferred regulatory control to rest with FDA and not the FTC as proposed in the Mead bill.[27] They were not pleased that their advertisers were balking at new contracts until the business of a new law was settled. They were not pleased that in the delay the number of bills touching advertising was mounting in Congressional halls—sixty by September.[28]

The food industry had always been more receptive to the various Copeland bills than their brother trades. These bills affected food products less. More recently they had won a significant victory on multi-grading. By the early spring Copeland reluctantly agreed to an amendment exempting established proprietary food products from disclosing a list of ingredients on labels.[29] Charles Dunn obtained an "informal agreement" with Walter Campbell that the concessions would stand and that the Senate committee report would specifically state that the bill did not apply to harmless trade puffing.[30] What disturbed Dunn and

26 *Ibid.*, testimony by Fischelis, 101.
27 "Composite Foods Bill," *PI* 172 (January 17, 1935), 108-109.
28 "60 Advertising Bills," *PI* 172 (September 12, 1935), 10.
29 Dunn, *Federal Food, Drug, and Cosmetic Act*, 401.
30 Dunn to Campbell, March 11, 1935, Correspondence, Box 564.

his Associated Grocery Manufacturers was less S. 5 than those provisions of the Mead bill which vested food advertising control with the FTC.[31] Certainly he wanted the Copeland bill put into law and soon. Surrender, the proprietary medicine industry called all this. Self-interest, pure and simple, their brother tradesmen shouted back.

So Senator Copeland was relatively happy. At the March public hearings representatives from the Advertising Federation of America and the National Publishers Association led the way in calling for the early passage of S. 5.[32] Charles Dunn, speaking for the grocery interests, proved one of the strongest defenders of the advertising provisions to testify.[33] The oral support was appropriately backed up in the publications of the new proponents.[34] Virtually all these bodies had some pet amendments they would like to see added, but none was presented in a crucial tone. The direction of things looked very good.

On March 22, S. 5 was reported out of committee. The day represented yet another victory for Copeland. In forcing the report he had beaten efforts by Senator Josiah Bailey to withdraw advertising control powers from FDA and by Senator Bennett Clark of Missouri to delay consideration for another week.[35] Copeland even got some assistance, if equivocal, from Franklin Roosevelt. Early in the Congressional session many proponents of S. 5 came to feel that passage would hinge on the attitude of the President.[36] Dunn, for one, personally called this need to the attention of Copeland, who had promised to urge the cause on FDR.[37] Urge was the correct word, for in January 1935

[31] *Ibid.*, February 27, 1935.
[32] *Senate Hearings on S. 5* (1935), testimony by Alfred T. Falk and Charles Parlin, 92, 56.
[33] "Food and Drug Harmony," *Business Week*, March 9, 1935, 9.
[34] "The Food Industry Approves S. 5," *Food Industries* 7 (April 1935), 162; Campbell to Sam Rayburn, June 7, 1935, Correspondence, Box 565; Falk to Campbell, January 17, 1935, *ibid.*, Box 564.
[35] "Food and Drug Bill to the Fore," *Business Week*, March 30, 1935, 7.
[36] "Food and Drug Bill Launched," *ibid.*, January 12, 1935, 19.
[37] Dunn to Campbell, February 27, 1935, Correspondence, Box 564.

the President denied he had given specific approval to S. 5. He told a press conference that he had not seen the measure but that he had doubts about whether it had "real teeth."[38]

By March, Copeland had talked with FDR along with Rexford Tugwell and Representative Sam Rayburn of Texas, chairman of the House Committee on Interstate and Foreign Commerce.[39] There were rumors afoot that the President was to make some public statement on drug legislation, though no one, including FDA, seemed to know what he would say.[40] On March 22 the message came. It was the same day, though apparently coincidental, that the Senate committee reported S. 5.[41] The brief message could have been stronger. FDR called for a new food law but, significantly to many observers, failed to mention the Copeland bill by name.[42] Still, a half loaf was better than none. The message received wide press coverage.[43] It made clear that a new statute had some priority in New Deal reform. It guaranteed that the Copeland bill would have a place on the present Congressional agenda.[44]

Surely it seemed that a new law was only weeks away. A good solid push from the consuming public and victory would be at hand. Unfortunately there was no rumbling at the grass roots. Perhaps the public was too overwhelmed by the depression to work up fervor over a drug bill. Perhaps drug reform was simply too dull by comparison with more spectacular developments of the New Deal to attract wide attention. Whatever the case the apathy was deadly. Copeland knew it. By mid-April he would say in despair,

[38] *Roosevelt Press Conferences*, January 16, 1935, Vol. 5, 54, Roosevelt Papers.

[39] Roosevelt to Marvin McIntyre, March 7, 1935, OF 375, Roosevelt Papers.

[40] Louis Engel to Lamb, February 6, 1935; Lamb to Engel, February 8, 1935, Correspondence, Box 569.

[41] "S. 5 Is before Senate," *PI* 170 (March 28, 1935), 46.

[42] *Ibid.*; "Food and Drug Bill to the Fore," *Business Week*, March 30, 1935, 7.

[43] Many press reports may be found in FDA Scrapbooks, Vol. 6.

[44] "Food and Drug Bill to the Fore," 7.

"if the American people do not want the bill what's the use of me wasting my time with it."[45]

The immediate gloom, however, derived not merely from public apathy. It owed much to the bitter criticism of important organizations that presumably spoke for the consumer. *Plain Talk,* bristling with an anger at FDA which went back to the ergot controversy, sharply condemned the agency's effort for a new bill as engineered by the American Medical Association. Walter Campbell might well wish the charges had had a bit of substance. The AMA was decidedly discontented with S. 5. For a time during 1934 it appeared that Copeland might have the backing of that very important body, but with 1935 the fallacy of the belief was clear. The AMA's Dr. William Woodward bluntly told the Senate committee that S. 5 had been so "whittled away and undermined in detail that you will not have an effective bill if S. 5 is enacted in its present form." *JAMA* admitted the measure had good points but called it "disappointing in its weakness."[46]

One state group, the Medical Society of New Jersey, was even demanding a Congressional investigation of alleged lax enforcement practices in FDA.[47] That charge was not typical of the medical profession and, in this case, was probably a carry-over from the ergot affair. The New Jersey group had been a center of the agitation over ergot and the home base of Dr. Edward Ill who had directed the report of the American Association of Obstetricians, Gynecologists, and Abdominal Surgeons.[48] The current

[45] *New York Journal of Commerce,* April 10, 1935, FDA Scrapbooks, Vol. 7.

[46] *Senate Hearings on S. 5* (1935), testimony by Woodward, 299; "Food, Drug, Therapeutic Device, and Cosmetic Legislation Pending in Congress," *JAMA* 105 (December 21, 1935), 2062.

[47] *Foods, Drugs and Cosmetics, Hearings before a Subcommittee of the Committee on Interstate and Foreign Commerce,* House of Representatives, 74th Cong., 1st Sess. on S. 5, held July 22, 24, 25, 31 and August 6-10, 12, 1935, testimony by Dr. Norman Burrit, 63. Hereafter cited as *House Hearings on S. 5* (1935).

[48] The circumstances of the report are examined above in chapter 1.

rhetoric smacked of Howard K. Ambruster, who was still actively waging his personal war on the drug administration. In 1935 his small book entitled *Why Not Enforce the Laws We Already Have?* was published. It was characteristically sensational. "Either he should be sued for libel," said Harry Elmer Barnes in reviewing the work, "or those who he denounces should be relentlessly exposed and properly punished." Ambruster had retained close connections with the New Jersey medical men.[49]

The charges out of that state were a minor irritation. The major one was that organized medicine was not pushing S. 5 to passage. The desire for a stronger bill was of course quite justified, but the AMA had not done battle for S. 1944 either. The women's organizations also wanted stronger provisions, but in the meantime they fought to keep what was left and to get what they could passed. The American Medical Association at least officially did little but object to losses. This matter was no small grievance to the women and especially to Mrs. Harvey Wiley who could remember what a vital role the AMA played in the passage of the 1906 statute. "They have opposed every bill at every hearing, even S. 1944 . . . ," she wrote to Mrs. W. D. Sporborg of the General Federation of Women's Clubs. "I can only surmise that the AMA and Consumers' Research do not want to yield their claim of superior knowledge to anyone else."[50]

This was a harsh criticism and surely one made in the anger of the moment. Yet the equivocal role of the nation's greatest medical group did provoke criticism. Their position was inappropriate for an organization which was charged with the national health and which understood well the power of pressure politics. The reasons for the reluctance to associate with any specific measure is hard to fathom, except perhaps as a residue of nervous self-

[49] USDA to Louis Howe, August 1935, OF 375, Roosevelt Papers; *New York Tribune* (November 17, 1935), FDA Scrapbooks, Vol. 9; "Active Ambruster," *D&C Industry* 35 (May 1935), 542.
[50] Wiley to Sporborg, February 19, 1935, Correspondence, Box 713.

interest from the 1920s. In that decade and the early 1930s the AMA labored against the crusade for compulsory health insurance.[51] The depression had brought popular criticism of medical costs. Under the circumstances, perhaps that organization was simply hesitant to encourage any governmental measure which touched upon the medical field.

Certainly AMA spokesmen were particularly touchy where S. 5 impinged upon their favored position in the field of American medicine. Dr. Woodward wanted in the Copeland measure specific exemptions from label requirements for the prescriptions of physicians. He wanted the Homeopathic Pharmacopeia dropped as a legal standard from S. 5. The logic for the request was good, but it did ring a bit of self-interest. Woodward also saw no reason to include trade representatives on the committee on health as established in the bill. The implication was that the committee should be in the hands of physicians—precisely what the drug industry feared.[52] Whatever the cause, the American Medical Association provided little help in the drive to get a new food-drug statute during 1935.

Then there were the militant consumer organizations. Benjamin Marsh of the People's Lobby appeared before the March hearings and flatly rejected Copeland's contention that S. 5 was the best bill he could expect to get into law. The Democrats controlled both houses, did they not? A stronger measure was possible. Marsh departed after a threat to take his grievances to Huey Long. Long could "make a speech on why the Democratic Party is afraid to protect the consumers . . . which would out-Farley Farley."[53] The journal *Consumers Defender* also trained its guns on the Copeland bill. The publishers wrote Roosevelt that the present bill was worse than existing legislation, and editorialized that S. 5 "has sedulously avoided . . . require-

[51] The best reference for this story is Burrow, *AMA: Voice of American Medicine.*

[52] *Senate Hearings on S. 5* (1935), testimony by Woodward, 165; also see *Senate Hearings on S. 2800* (1934), testimony by Woodward, 373.

[53] *Senate Hearings on S. 5* (1935), testimony by Marsh, 143.

ments for adequate food and drug legislation." For that periodical, "must" additions were: complete formula disclosure, federal licensing of medicine manufacturers, multiple grades on food, and an outright ban on drug advertising for diseases where self-medication was dangerous.[54] In particular there were the powerful Consumers' Research and "America's No. 1 professional consumer," Arthur Kallet. He showed up at the March hearings angry and bearing fifteen pounds of documents.[55] For Kallet, S. 5 was totally inadequate. He lambasted the bill, Copeland, and the committee, announcing that, "as to what I propose . . . I do not expect it to receive any consideration by this committee or by Congress." His diatribe was halted by the committee after fifty minutes but not before he accused committee members of acting like "American prototypes of Hitler." CR's *Bulletin* continued to lash out at S. 5 and "the brazen contempt of the Roosevelt administration for the welfare of the consumer [which] has again placed Senator Royal S. Copeland, employee of quack medicine manufacturers, in charge of food, drug, and cosmetic legislation." CR condemned also the women's organizations for "humbly accepting the few crumbs . . . dropped into their laps."[56] As for Copeland, he remained philosophical about the personal attacks even in the dark days of the summer. "I should rather have all men speak well of me," he wrote Swann Harding, "but when I know some men won't, why think about them?"[57] The thought was good, but the Senator surely knew that Consumers' Research was no help in getting S. 5 on the law books. Copeland needed consumer support—any consumer support. He might well have given grim concurrence to a recent comment by *Food Industries* on Arthur Kallet: "After all, the professional consumer lives on issues—not

[54] March 1935, FDA Scrapbooks, Vol. 6.
[55] "The Talk of the Industry," *Food Industries* 7 (April 1935), 157.
[56] *Senate Hearings on S. 5* (1935), testimony by Kallet, 69, 64-65, 75; *General Bulletin of Consumers' Research*, 1, n.s. (April 1935), 11.
[57] Royal S. Copeland to Harding, July 12, 1935, Harding Papers.

their settlement, for a settled issue is a dead issue and people do not pay their money to maintain fighters for a cause that no longer exists."[58]

While CR and other militant groups continued to speak to the bill "which is not," the bill "which is" almost passed into oblivion. By the end of March, S. 5 had reached its high-water mark for the year. Copeland still had some advantages in his struggle, not the least of which was some strong support for his legislative efforts within the affected industries. What he did not have was the backing of the proprietary drug men. By April they stood almost alone in their adamant opposition to the Senator's measure, but theirs was a powerful voice. It was likely that even Copeland did not realize how powerful—as yet. There were, however, those who did understand. "If I were Senator Copeland," one very astute observer was heard to comment at the March hearings, "I would rather the opposition of everybody in this room and have the support of Frank Blair [Proprietary Association] than to have it the other way around."[59]

Nor could the New Yorker realize that at the seemingly bland March hearings the major lines of the food bill battle had been laid out for the next three years. At the time, Copeland could not see these things. He did know that the medicine industry wanted more elaborate court review provisions than were provided in S. 5, but that was one of many changes which they desired.[60] There seemed nothing ominous in the appearance at the hearings of Samuel Fraser, speaking for the International Apple Association. Fraser also wanted to extend the court review sections. He was concerned about the old question of spray residue on fruit and the power of the Secretary to establish residue tolerance levels which carried the force of law. The apple producers wanted the right to contest criteria of the Sec-

[58] "The Talk of the Industry," 158.
[59] "Drug Bill Goes to Hearings," *D&C Industry* 35 (March 1935), 266.
[60] Ole Salthe to Copeland, April 16, 1935, Correspondence, Box 565.

retary's decision in the courts, not just to contest whether their apples exceeded the established tolerance levels.[61] To Copeland, Fraser must have been simply one more critic. He could not see that court review would be the crucial question in 1938.

What sounded of more significance at the March hearings was the matter of which agency would regulate food, drug, and cosmetic advertising. It was not foreseeable, however, that this question would prove to be the insoluble battle line of 1936. The move to strip FDA of proposed advertising authority, leaving all advertising regulation with the Federal Trade Commission, was spearheaded by the Proprietary Association and the Institute of Medicine Manufacturers.[62] To *Business Week* the reason for the drive was obvious. The previous year the medicine manufacturers had smelled the possibility of victory and fought successfully to block the passage of any new bill. This year their belligerent self-assurance was waning. They were seeking to make sure that if a bill passed it would be as ineffective as possible. Because of the cumbersome cease and desist procedure under which FTC operated, as well as such restrictive judicial precedents as the Raladam decision, according to the *New Republic*, the nostrum vendors were sure they could triumph in court no matter how bad their advertising was.[63]

The *New Republic* exaggerated the judicial confidence of the medicine trade. FTC, even with the Raladam albatross, was not totally ineffectual, but certainly the cease and desist procedure was slow. The obstacles to successful application were many. The risks of misleading advertising were also much less under FTC than under the criminal-prosecution procedures open to FDA under S. 5, since action by FTC brought no sanction, merely the necessity of halting the practices complained of. Some nostrum pro-

[61] *Senate Hearings on S. 5* (1935), testimony by Fraser, 255-69.
[62] "Too Much Commission," *PI* 170 (March 14, 1935), 108; *Senate Hearings on S. 5* (1935), testimony by James Hoge, 133-34.
[63] "Beware of the Medicine Men," *New Republic* 82 (March 6, 1935), 90.

moters dared to dream even further. The *Oil, Paint and Drug Reporter* suggested editorially that the trade should consider backing some additional power for the Federal Trade Commission. If this were done it might be possible to convince Congress that there was no longer any need to pass a new food and drug act.[64]

Needless to say, FTC's trade backing expressed no such motives to members of Congress. At the March hearings and after, manufacturers simply lined up solidly to defend the Commission's honor and efficiency. They lauded the effectiveness of the cease and desist procedure and the appropriate business orientation of FTC.[65] Robert Fischelis, President of the American Pharmaceutical Association, was shocked at the illogic of such talk. The primary concern of the Federal Trade Commission was trade promotion. "It would be a calamity," he told the APhA convention in 1935, "if a branch of the Federal Government whose chief function is trade promotion should be entrusted with the control of the advertising of medicine."[66] *Printers' Ink* reacted in a less analytical way. How very curious it was, the journal mused, to see businessmen who not long before saw FTC as a "nosy nuisance" now "throwing themselves into the Commission's arms." As for the rhetoric behind the move, that was simply too much for *PI*. "To say that the 'Commission has done a good job with advertising for years' is, to put it mildly, being deliciously naive. The Commission has of course done nothing of the kind."[67]

Most of the food and advertising industry as well as segments of the cosmetic and drug trade opposed Commission jurisdiction. They preferred to have their ad copy in the hands of FDA, whose personnel had a scientific background. More important, they wanted a law put on the

[64] February 18, 1935, FDA Scrapbooks, Vol. 5.

[65] "Observations of the Day," *Standard Remedies* 21 (April 1935), 1; "Too Much Commission," 108; *Senate Hearings on S. 5* (1935), testimony by James Hoge, 133-34.

[66] "Address of the President of the American Pharmaceutical Association," *JAPhA* 24 (August 1935), 638.

[67] "Too Much Commission," *PI* 170 (March 14, 1935), 108.

books soon, and they saw no logical reason to make an issue over where advertising control would rest.[68] Because of this business indifference to the case for FTC the whole matter might have been soon resolved, if the conflict had remained one of Copeland versus the nostrum men. Proprietary backers of the Commission would have been unable, very probably, to enlist sufficient Congressional support to press their case. At best, these interests might have gained a few more concessions in some type of compromise with Copeland.

Compromise was a viable possibility. The New York Senator's chief assistant Ole Salthe was working for it by spring and believed something could be accomplished. After a long conference with Frank Blair and James Hoge of the Proprietary Association he conveyed his hopes to Copeland. First of all, Salthe reported, Blair and Hoge were resentful about the treatment they had received compared to that given Charles Dunn. Dunn had managed to get his own "jokers" incorporated in the bill and was still viewed as a champion of food and drug reform. Blair and Hoge had tried to do the same and had come off villains. Salthe believed that with a bit of mollifying treatment the two men would be willing to compromise their FTC demands. Their real grievance was procedure to be followed against advertising which was held to be objectionable. A cease and desist order was better than being made to look like a criminal through FDA judicial action. Salthe was sure that Hoge and Blair would abandon the Commission if some type of court review proceedings, prior to criminal prosecution, could be incorporated in S. 5.[69]

Thus some agreement was possible. For that matter, Copeland was able to push his measure through the Senate without any serious concessions. The crucial nature of the

[68] *Senate Hearings on S. 5* (1935), testimony by Ernest Little and Hugo Mock, 191, 197; "Composite Foods Bill," *PI* 170 (January 17, 1935), 109; "S. 5 Advertising Section," *ibid.*, March 7, 1935, 52; "Food and Drug Harmony," *Business Week*, March 9, 1935, 9.

[69] Salthe to Copeland, April 16, 1935, Correspondence, Box 565.

advertising control issue in 1936 came because Copeland's bill ran against the interest of the Federal Trade Commission itself. The Commission had powerful Congressional friends, particularly in the House, and FTC was quite adroit in the use of its influence. Compromise with the medicine trade would be only half the battle.

This bureaucratic tangle was beginning to materialize by the time hearings on S. 5 were held. Ewin Davis, Chief Commissioner of FTC and a former member of the House, had voiced his objections to the division of advertising responsibility at the hearings on S. 2800. He had defended the operation of his agency and down-rated the consequences of the Raladam court decision. He had warned of the waste, confusion, and duplication of services if two agencies were charged with advertising control. Davis was back to restate his case at the S. 5 hearings, and to go one step further. If it were developed that past court decisions had hampered FTC, the Commissioner stated, "we respectfully suggest consideration of amending the Federal Trade Commission Act, instead of depriving them of jurisdiction in order to give it to another agency."[70] In December 1934 Senator Burton Wheeler introduced Senate bill 944 to strengthen the power of FTC. It was not a threat to the food bill yet, but the day was soon to come.[71]

In the meantime, and while Davis emphatically professed to leave the decision on advertising in the able hands of Congress, the Commission began a quiet campaign to subvert the advertising clauses of S. 5. Davis pressed the cause on Franklin Roosevelt. Senator Josiah Bailey received a visit from Commissioner Garland Ferguson which was followed up by a letter from Ferguson listing the trade groups lined up with FTC. The Commission was at work also in the House.[72] So the process went, much to the irritation of

[70] *Senate Hearings on S. 5* (1935), testimony by Davis, 115.

[71] For the complete history of this legislation see Charles W. Dunn, *The Wheeler-Lea Act: A Statement of Its Legislative Record* (New York, 1938).

[72] Roosevelt to Marvin McIntyre, February 9, 1935, OF 375, Roosevelt Papers; Ferguson to Bailey, February 26, 1935, Bailey

FDA's proponents, including Rexford Tugwell. The Assistant Secretary expressed his resentment in a letter to Donald Richberg, director of the National Emergency Council early in the year. Perhaps remembering past troubles over the Deficiency Appropriations Act of 1919, he pointed out that government agencies were prohibited from lobbying their cause before Congress. FTC should be instructed to stop pressing their views on S. 5.[73]

Tugwell believed the decision on advertising should rest with Roosevelt.[74] The Secretary and Davis met with the President to discuss the matter on February 14. According to FDA's Charles Crawford, Roosevelt then told both men that the bureaucratic "squabbling" must cease and that he would be guided in this issue by the recommendations of Senator Copeland.[75] That sounded good, but in fact a basic point in the future gravity of the jurisdictional dispute was that FDR never made a clear decision. He remained equivocal and allowed settlement to pass to Congress without guidelines. The advertising crisis was in the future, however. For the time being, it was troublesome, very troublesome, but Copeland could see himself as coming out on top.

The more immediate problem and the crucial one for the spring of 1935 was the power of FDA to make multiple seizures of goods. The March hearings were one of several occasions upon which medicine representatives in particular had voiced objection to that power. This time their grievances set the stage for a showdown. For Copeland the reality of the showdown would become apparent when S. 5 came to the Senate floor for debate. He had forced the bill out of committee over the objections of Senators Josiah Bailey and Bennett Clark. Neither man was enthusiastic about the measure, a fact some observers attributed to the

---

Papers; Otis Pease, *The Responsibilities of American Advertising* (New Haven, 1958), 121-22.

[73] Rexford G. Tugwell to Richberg, January 24, 1935, Correspondence, Box 564.

[74] *Ibid.*

[75] Roosevelt to Marvin McIntyre, February 9, 1935, OF 375, Roosevelt Papers; Crawford to Cavers, February 14, 1935, Commissioners' File, Box 11.

influence of the Vick Chemical Company in Bailey's North Carolina and Lambert Pharmaceutical Company, of Listerine fame, in Clark's Missouri.[76]

On April 1, 1935, S. 5 came up for debate. Copeland hoped for quick passage, but that was not to be. Bailey had "certain" amendments ready, as he explained to a constituent who had written on behalf of Vick.[77] With the help of a few colleagues the North Carolina Senator and Clark kept S. 5 on the floor for eight days. The purpose of this strategy was apparently to provide time to build support. Copeland understood this and twice challenged Clark to call for a vote on his motion to recommit the bill to committee. The canny Senator from Missouri refused, fearing that a defeat would hurt his cause. On the fourth day of debate the Senate approved the filing of a minority report from the Committee on Commerce. It was signed by Bailey, Clark, Joseph Guffey, N. L. Bachman, and A. V. Donahey.[78]

The text of the report summed up the basis of the Bailey-Clark attack. Advertising was properly the bailiwick of FTC. Section 401 included a statement defining adulterated drugs as those dangerous to health under conditions of use as prescribed in labeling or advertising. This statement should be shifted to the misbranding section of the bill. The shift was extremely important because adulterated drugs were subject to multiple seizure. That would not be the case with misbranded goods if Bailey had his way. On misbranding, FDA would be limited to one seizure except upon *proof* by the Secretary that said goods were imminently dangerous to health. Finally, the minority held that judicial proceedings on the single seizure should upon motion be removed to a jurisdiction of reasonable proximity to the residence of the claimant.[79]

[76] "Solemn Act," *Time* 25 (April 15, 1935), 13; David F. Cavers, "The Federal Food, Drug and Cosmetic Act of 1938," *Law and Contemporary Problems* 6 (Winter 1939), 13.

[77] Herbert Peele to Bailey, March 28, 1935, Bailey Papers.

[78] The report is reproduced in Dunn, *Federal Food, Drug, and Cosmetic Act,* 389-400.

[79] *Ibid.*; Cavers, "Federal Food, Drug, and Cosmetics Act," 15-16; James Hoge, "Multiple Seizures," *PI* 171 (April 18, 1935), 58-62.

These suggestions, with the exception of FTC control on advertising, were incorporated into the so-called Bailey amendment which came up for a vote on April 8. To accept the amendment would mean a very serious blow to drug controls. Proprietary drugs would be virtually immune to multiple seizure, a procedure which had been FDA's most effective weapon in the past. The danger in patent medicine was seldom a lethal ingredient. The danger lay rather in encouraging a sufferer to use some worthless proprietary balm when his safety was dependent upon prompt treatment with an appropriate remedy. It was the danger, for example, of using the celebrated horsetail weed extract, Banbar, in lieu of insulin for treatment of diabetes. The Bailey amendment, according to Ruth Lamb, would allow colored tap water to go on sale as a cure for cancer. FDA would be restricted in such a case to one seizure unless the manufacturer added arsenic to each dose, making the contents "imminently dangerous."[80]

Unfortunately for Copeland and FDA, few people saw this complex and confusing issue as quite that crucial. Except for the women's organizations there was no wrathful reaction from the public. In the Senate the New Yorker reasoned, cajoled, and threatened his colleagues with his own withdrawal from sponsorship of S. 5.[81] The effort was highly unrewarding. It was even hard to keep an audience. On one occasion the apathy reached comic opera proportions. Private conversations on the floor made it difficult to hear Copeland's speech. The Sergeant at Arms was called to restore order, but he was not to be found. The Senators ceased their conversations only because their attention was attracted to a greater hubbub in the gallery. It was ordered that the galleries be cleared. That took time, but at last Copeland got up to resume his speech. Alas, there were

[80] Cavers, 15-16; Lamb to James Rorty, November 19, 1935, Correspondence, Box 569.

[81] For a convenient statement of the full debate see Dunn, *Federal Food, Drug, and Cosmetic Act*, 269-476. For the crux of Copeland's statements see *Congressional Record*, 74th Cong., 1st Sess. (April 5, 1935), 5137-40.

few to listen. With sorrow he observed "we seemed to have cleared not only the galleries but the floor. I suggest the absence of a quorum."[82]

The Bailey amendment was carried on April 8. Both sides understood the vote as a test case. If the amendment failed, Copeland would immediately push for a final vote of the whole bill. In the event the amendment passed, Clark would press his motion to recommit the measure to committee, where advertising control provisions would be stripped.[83] Only the timely intervention of majority leader Joe Robinson prevented the latter. Robinson, a proponent of drug reform, brought up the District of Columbia appropriations bill which had floor right-of-way and thus allowed S. 5 to return to the calendar. There, the future of the bill was problematical. It was safer than in committee, but it was unlikely to again reach the Senate floor during the session unless brought up out of sequence by a majority vote.[84]

The American Pharmaceutical Association's *Journal* summed up well the reaction to this legislative intrigue of those in the food law reform camp: "It is difficult to understand why minority groups, whose chief interest in food and drug legislation is . . . to ease the path of fakers in the food and drug industries, should be in a position to thwart the efforts of earnest legislators and respectable citizens in providing proper control of the manufacture and distribution of drug products."[85] Senator Copeland's reaction was also anger, but anger with a purpose. "As far as I am concerned," he told the press, "it [S. 5] is dead for this session." "We'll see whether the devil wins or the forces of righteousness prevail."[86] The strategy was clear to everyone. Cope-

[82] "Solemn Act," *Time* 25 (April 15, 1935), 13-14. For the full record of this episode see *Congressional Record*, 74th Cong., 1st Sess. (April 2, 1935), 4840-51, 4858-69.
[83] "Food and Drug Fight Checked," *Business Week*, April 13, 1935, 16; "S. 5 Again Sidetracked," *PI* 171 (April 11, 1935), 28ff.
[84] *Ibid.*, 91.
[85] Editorial, *JAPhA* 24 (May 1935), 349.
[86] *New York Journal of Commerce*, April 10, 1935; *New York Produce Guide*, April 13, 1935, FDA Scrapbooks, Vol 7.

land hoped to counterbalance his loss on the Senate floor by provoking a strong popular reaction. If the reaction came he might swing over enough of his colleagues—for many were less enemies of drug reform than apathetic to the issue—to eliminate the Bailey amendment. The plan was not bad. Grass roots reaction was hardly spectacular. By and large it was limited to the women's organizations, but the women were angry.[87] That anger was a start.

Meanwhile, as Copeland sat back, Clark and Bailey began to get more than a bit nervous. Charles Crawford wrote David Cavers that both men were professing to be "outraged" that Copeland refused to push the bill further. Crawford was concerned about the matter and seriously doubted that the two men were really "outraged." He believed they were preparing to introduce their own substitute bill. It would undoubtedly be the old Mead bill from the House. If such a measure passed and the President signed it, Crawford bemoaned, the "patent medicine crowd" would have "a holiday for an indefinite number of years."[88]

Crawford was overly pessimistic. In all likelihood Bailey and Clark were concerned, very concerned. It was a matter of practical politics. There was little chance that the Mead bill could be put through the upper chamber. If nothing else, Copeland's presence as committee chairman and the tradition of Senatorial courtesy were tremendous obstacles. On the other hand, if the New Yorker made no further move and S. 5 died, Clark and Bailey would be singled out for blame. Women's organizations were unpleasant foes. Moreover, the bill was an Administration measure. Roosevelt was not always ardent in his support, but the current stalemate followed right on the heels of a Presidential request to pass a new law. It was this matter which seemingly proved decisive.

By April 17, Bailey had become concerned enough to write Roosevelt. The Senator wished to explain his reasons

[87] "S. 5 Again Sidetracked," *PI* 91; "Drug Bill Revival," *Business Week*, May 25, 1935, 22.
[88] Crawford to Cavers, April 10, 1935, Commissioners' File, Box 10.

for amendment. He, like FDR, wanted "a first class act" and one in full accord with the President's message. The credit for the new law, Bailey continued, should be Copeland's. The North Carolina Senator did not desire to introduce a substitute because it would detract from Copeland's efforts.[89] Roosevelt passed the letter to Rexford Tugwell for consideration. Long unhappy with the weakened provisions of S. 5 in general, the now Under Secretary wrote FDR that the Bailey amendment was "emasculating." The public would have virtually no protection against innocuous but useless drugs. Bailey's contention that his amendment did not materially alter seizure powers provided for in the 1906 statute was "misleading." That law had no restrictions on multiple seizures.[90]

Tugwell also prepared a draft reply to Bailey for the President's signature which was sent as drafted. The letter was not commanding in tone but made it clear that FDR wanted a new law which would be "comprehensive in its terms and vigorous in its enforcement provisions." It also stated that the President would not like to see seizure provisions "so curtailed as to prevent summary and unrestricted action" on articles harmful to health.[91] The letter was Tugwell's, and it is impossible to ascertain how strong the President's personal feelings on the matter really were. What is clear was that the personal attention of the Chief Executive could be a powerful force. Bailey replied promptly, again assuring the President of his support for a new food and drug bill. His amendment had been offered only because the seizure provisions in S. 5 were "foggy" and not clear. "I do not think," he concluded, "there is any real difference between us on the matter."[92]

Roosevelt met personally with Bailey on May 10, but

[89] Bailey to Roosevelt, April 17, 1935, OF 375, Roosevelt Papers. Also see some correspondence in Bailey Papers.

[90] Tugwell to Roosevelt, April 24, 1935; Roosevelt to Bailey, April 26, 1935, OF 375, Roosevelt Papers. On the latter, also see Bailey Papers.

[91] Ibid.

[92] Bailey to Roosevelt, May 1, 1935, OF 375, Roosevelt Papers.

many days before that conference negotiations were under way between the North Carolina Senator and Copeland. By May 16, a compromise was complete.[93] The original amendment restricted FDA to one seizure except upon "showing by the Secretary" that an article was misbranded so as to be "imminently dangerous" to health. The compromise version allowed multiple seizure when the Secretary "has probable cause to believe from the facts" that an item was imminently dangerous. In the revised Senate committee report Copeland was careful to explain that it was intended that multiple seizure on misbranded goods was not confined to poisonous goods. It would include innocuous drugs with false claims for a serious disease which might induce the patient to delay effective treatment.[94] The shift of the "conditions of use" statement from the section on adulterations to the section on misbranding remained in the new version. The change of venue provision also was retained.

The compromise was better than the original Bailey proposition, but it still represented a significant victory for opponents of a strong law. The 1906 statute contained no such restrictions on multiple seizure. FDA was not happy with the compromise and hoped to strengthen the seizure clause before final passage.[95] The women's organizations felt the same.[96] For the moment, however, both had decided to live with it. Without heavy public support there was little else they could do, and that support was not forthcoming. In June, the periodical *Survey* was graphic about the absence. There was little to relate about the bill "except the apparently fatal lethargy of the consuming public. There have been no effective public protests against

[93] *Ibid.*; *OP&D Reporter*, May 20, 1935, FDA Scrapbooks, Vol. 7.
[94] For convenient comparison of phrasing in the matter of seizure actions see Dunn, *Federal Food, Drug, and Cosmetic Act*, 467, 528. The revised Senate report is also reproduced in Dunn, 477-90.
[95] Lamb to Rorty, November 19, 1935, Correspondence, Box 569.
[96] *Christian Science Monitor*, May 28, 1935, FDA Scrapbooks, Vol. 7.

existing and proposed amendments which weaken the measure so seriously."[97]

The reaction of the medicine industry was mixed. Some segments disliked the compromise, but on the whole the trade was pleased. A special bulletin of the Proprietary Association announced that that organization's major objections had been met, and they no longer opposed passage.[98] The nostrum trade had reason to be pleased. *Drug and Cosmetic Industry* made the point: Copeland and Campbell must now realize that they cannot get a bill by without the aid of the drug people. ". . . the drug industry alone was able to stop them."[99] On May 28, 1935, Senate bill 5 was brought up and passed by the upper house without opposition. The battle had been long and not totally satisfying in result. Perhaps, just perhaps, the nation would have a new food and drug statute very soon. It was up to the House—the House, "aye, there's the rub."

[97] "Ferment in Washington," *Survey* 71 (June 1935), 176.

[98] Proprietary Association, *Special Bulletin*, May 29, 1935, Correspondence, Box 565; *New York Journal of Commerce*, June 7, 1935, FDA Scrapbooks, Vol. 7.

[99] "Maybe a Drug Act," *D&C Industry* 36 (May 1935), 542.

V

"TUGWELL AND TENNESSEE
BEAT US"

Now, members of the House, what are you going to
do about it? Are you going to turn this over to
Tugwell for enforcement or are you going to leave
it with the Federal Trade Commission with such men
as Judge Davis and other men from this House on
the Commission?

REPRESENTATIVE SAMUEL MC REYNOLDS
HOUSE DEBATE ON SENATE BILL 5
JUNE 1936

THE intrigues of food and drug legislation had become
old hat to Royal Copeland by 1935. He was experienced
and battle-worn. For Virgil Chapman, Representative from
Kentucky, the summer of 1935 was a shocking revelation.
Copeland and Walter Campbell had hoped to have S. 5
proceed through the House without further public hearings. They had even obtained the backing of the Proprietary Association's James Hoge for this move. Unfortunately,
Representative Sam Rayburn, Chairman of the House Committee on Interstate and Foreign Commerce, felt otherwise. There must be new hearings.[1] Thus Virgil Chapman
entered the field of the food and drug battle. He was
chosen to chair the new hearings. "In the three weeks the
committee had sat, more startling information of direct
and immediate consequences to the public . . . had been
revealed," the Kentucky Congressman purportedly stated
later, "than before any other committee throughout the
life of this Congress."[2]

[1] Walter Campbell to Ole Salthe, June 7, 1935, Correspondence,
Box 565. Ole Salthe to Campbell, May 29, 1935, *ibid.*
[2] Charles Crawford to David F. Cavers, August 8, 1935, Commissioners' File, Box 11.

There were the expected things. Representatives of the various women's organizations appeared in support of S. 5 and continued their demand for the restoration of quality grade labeling for processed food.[3] A spokesman for Consumers' Research delivered that organization's usual criticism of Copeland's bills.[4] William Woodward of the American Medical Association again took an equivocal position. At the hearings on S. 2800 he had objected to the inclusion of the Homeopathic Pharmacopeia in the bill as a drug standard. He asked now that all standards be removed since, he believed, the recent Schechter poultry case denied Congress the right to delegate power to make binding regulations.[5] His answer was to create a federal pharmacopeia.[6] Walter Campbell was present to ask that seizure provisions be strengthened and to plead FDA's need for advertising control powers.[7]

What made the hearings special to Chapman was his own gradual recognition of the intricacies involved in FDA's current struggle for a more adequate law. The need for revision was brought home by such witnesses as Mrs. Alvin Barber representing the American Association of University Women. Mrs. Barber was late for her turn before the committee because she had gone out between sessions to see if she could purchase a package of Lash Lure, an eyelash dye which had blinded some of its users. Mrs. Barber found four shops within six blocks where the concoction was being used or was for sale in packages.[8] As a cosmetic, Lash Lure was beyond the scope of the present law. At the end of the hearings Chapman and several of

[3] *House Hearings on S. 5* (1935), testimony by Mrs. Harvey W. Wiley, 167; testimony by Mrs. Harris Baldwin, 373.

[4] *Ibid.*, testimony by J. B. Matthews, 507.

[5] For a summary of the case see "Schechter Poultry v. United States," in Carl Swisher, *Historic Decisions of the Supreme Court* (New York, 1958), 144-50.

[6] *House Hearings on S. 5* (1935), testimony by Woodward, 304-305.

[7] "The Course of the Drug Bill," *D&C Industry* 37 (August 1935), 174; *House Hearings on S. 5* (1935), testimony by Campbell, 80-84.

[8] *Ibid.*, testimony by Barber, 385.

his colleagues visited the Chamber of Horrors exhibit at the new FDA offices.[9] The need for a new law was further made clear.

As the hearings progressed Chapman began to probe with verve the devious nature of the opposition to new legislation. He pressed the matter of "pink slips" with Hugh Obear, attorney for the Lydia Pinkham Medicine Company. The Kentucky Representative asked if the Pinkham Company had sent out letters to the public calling for the defeat of S. 5. Obear answered in the negative. Chapman produced a so-called pink slip sent under the Pinkham name. Obear disclaimed any knowledge of it, but the Kentuckian was not persuaded.[10] What was Lydia Pinkham's Compound good for, the Kentucky Representative queried his nervous prey. Obear assured him of the nostrum's medicinal benefit. Wouldn't a "good drink" of Kentucky whiskey or even "moonshine" from Tennessee do just as much good as "this famous product," Chapman asked, adding that he knew one woman who was "half drunk" all the time on "some kind of tonic or another."[11]

The Blue Grass State Congressman also turned his lance on William Jacobs, Vice President of the Institute of Medicine Manufacturers. The topic was the "red clause" and letters which threatened to cancel advertising commitments if a new bill passed. Jacobs equivocated, squirmed, and at length was forced to acknowledge his awareness of such advertising threats. In the end the nostrum maker admitted having sent letters to southern newspapers warning of revenue loss in the event of a new law. He did not oppose a new law per se, Jacobs insisted defensively, only a number of features in S. 5. Chapman was disgusted. If the Institute representative did simply oppose any new law, the Kentuckian commented sarcastically, "you would be about the first such man . . . in the last 3 or 4 weeks. Everyone who has been here so far . . . has said, 'I am for this bill.' They

[9] *Drug Trade News*, August 19, 1935, FDA Scrapbooks, Vol. 8.
[10] *House Hearings on* S. 5 (1935), testimony by Obear, 241-44.
[11] *Ibid.*, 245.

have started with that and ended with, 'if this is put in,' or, 'if this is taken out.' "[12]

So the hearings went. By the end Chapman was dismayed. He intended to take to the radio and newsreels to garner support for a new law. "The public has no conception of the necessity for new food and drug legislation," he stated, "simply because they do not know what it is all about." In particular, Chapman felt that the press had not given the hearings the coverage they deserved.[13] More generally, he was concerned about the immediate future of the food-drug bill. Prior to the hearings Chapman hoped to push the measure to the floor for a vote before the House adjourned. By the end he changed his mind. The hearings had been all too "pat." The Kentuckian "was astute enough," Charles Crawford wrote David Cavers, "to smell a mouse in the unanimity and fervor with which the big patent medicine interests . . . recommended the passage of the bill."[14]

While many trade representatives had called for the immediate passage of S. 5, too many more had changes to offer. In particular the hearings gave indications of some strong sentiment in the proprietary medicine industry to leave sole control of advertising with the Federal Trade Commission.[15] When Chapman realized how controversial were the issues of amendments he concluded it would be best not to attempt to rush the bill through. To do so might well provoke a floor fight in the House which Chapman was not sure he could win. S. 5 could be stripped of its advertising control provisions and otherwise emasculated while Chapman stood by helplessly. He determined that the best strategy was to hold the bill in committee and at-

[12] *Ibid.*, testimony by Jacobs, 681-82.

[13] Crawford to Cavers, August 8, 1935, Commissioners' File, Box 11; *Food Field Reporter*, September 23, 1935, FDA Scrapbooks, Vol. 8.

[14] Crawford to Cavers, August 8, 1935, Commissioners' File, Box 11.

[15] *House Hearings on S. 5* (1935), testimony by Hugh Craig, 361; Dr. Edwin Newcomb, 458; Clinton Robb, 68-70.

tempt a quiet "education" program among his colleagues until the necessary support for passage was assured. In the meantime the Kentuckian considered that it would be bad tactics even to prepare a committee draft of the measure which could only operate as a target for the opposition.[16]

Copeland's bill would rest quietly in the House committee for the next nine months, only to be reported out in May 1936. Some trade elements were bitterly angry about this matter and even sought abortively to force the measure from committee by petition.[17] Rumors of Chapman's motives were hot and heavy. One of the most popular was that the Blue Grass Representative was deliberately holding up the bill until Roosevelt yielded on certain matters of political patronage.[18] The validity of the rumor seems very doubtful. In the light of the President's lukewarm interest in Copeland's opus the measure would hardly constitute an effective political lever. Whatever the case, S. 5 was off the battleground of the Congressional halls for the time being. The center of combat became the public "front."

In the fall of 1935 part of the struggle for Walter Campbell was a certain "thunder on the left" in the person of Rexford Tugwell. In recent months much of Tugwell's time had been occupied with the activities of the Resettlement Administration which he now headed. Rumor had it, however, that "Terrible Rex" was about to throw himself back into the fight for a new statute. He was thoroughly disgusted with the weakness of the current bill and intended to demand the enactment of the original Senate version, S. 1944. Campbell threatened to resign, according to the story, if Tugwell made this move. The drug unit chief denied the whole affair, but the denial is hard to accept.[19] The rumors were too prevalent, too concrete, and indeed too logical.

[16] Crawford to Cavers, August 8, 1935, Commissioners' File, Box 11; *Drug Trade News*, February 17, 1936, FDA Scrapbooks, Vol. 10.
[17] *Ibid.*
[18] *Ibid.*
[19] *Ibid.*, November 11, 1935, FDA Scrapbooks, Vol. 8; *Kiplinger Agriculture Letter*, November 16, 1935, *ibid.*, Vol. 9.

A new drug statute was important to Tugwell, and he had been disgusted at successive "watering down" of provisions encompassed in the 1933 bill. If Tugwell had decided to make a fight for S. 1944, then the rumor of Campbell's reaction was consistent. Campbell was too much the realist to renew what was bound to be a fruitless battle. S. 5 might be weaker than S. 1944, but it had a significant amount of trade support. This support would probably be lost if Tugwell had his way. Plainly no bill could pass Congress in the face of unified opposition from the affected industries. Beyond the practicalities involved, Campbell believed a reversal of this sort would be taken, and quite correctly taken, by trade support as a breach of faith. He would have no part in this. Apparently the issue was settled by mid-November. Tugwell backed down, so related the *Kiplinger Agriculture Letter*, after Secretary of Agriculture Henry Wallace sided with Campbell.[20]

Yet it was also true that Walter Campbell was becoming restive about the slow progress of drug reform and the inability of FDA to take part in the fray. He was becoming less cautious about the supposed "lobbying" activities within his agency. With the death of S. 2800 in the summer of 1934, some slight and unceremonious change actually began. During the interval in which Congress was not in session Campbell authorized a renewal of speeches by agency personnel on the shortcomings of the 1906 statute. He did make clear, however, that speakers should avoid arguing *directly* in favor of a new law.[21] Presumably these activities were again suspended with the opening of Congress or at least held to such a minimum as not to attract the notice of the trade.

When Congress adjourned in the summer of 1935 the public "education" work of FDA resumed. The old Horrors exhibit of 1933 was no longer available to the public, but a similar display on the weakness of the 1906 law was avail-

[20] *Ibid.*
[21] Campbell to Chief, Western District, July 19, 1934, Correspondence, Box 437.

able in the form of photographs and lantern slides. FDA was less hesitant than before about accommodating public requests for the material. Indeed the drug unit began to accept, and seemingly to solicit quietly, dates for the displays at professional meetings.[22] Charles Crawford's correspondence with Lawrence Gourley of the American Osteopathic Association was a case in point. Gourley notified Crawford in February 1936 that in line with "our conversation" FDA had been allotted twenty-one square feet of exhibition wall space and eight square feet of table space at the summer convention of the Osteopathic Association.[23]

Another sign of Campbell's "loosened" attitude toward activities in the agency was the 1936 publication of Ruth Lamb's book, *American Chamber of Horrors*. The work was not officially sponsored by FDA. Miss Lamb published it as a private citizen, but Miss Lamb was FDA Information Officer. The volume was a devastating exposé of abuse in the food, drug, and cosmetic industries, and it was a clarion call for public support to pass a new statute. *Drug and Cosmetic Industry* was correct in its angry charge that even while the publication was not sanctioned by FDA, the book could never have gone to press without the support and approval of agency officialdom.[24] The Food and Drug Administration was becoming more aggressive.

Private groups concerned with consumers, principally the women, were doing the same. There was, of course, the now characteristic dissension in these ranks. During February the *American Journal of Public Health* labeled S. 5 as, "innocuous as the most resourceful drug manufacturer, advertiser, and salesman could desire." Either the bill must be significantly strengthened, or it should be voted down "as unworthy."[25] In May a spokesman for *JAMA* expressed the discontent of the American Medical

[22] Ruth Lamb to Benjamin Seidner, January 22, 1936, *ibid.*, Box 713.
[23] Gourley to Crawford, February 17, 1936, *ibid.*, Box 714.
[24] *D&C Industry*, March 1936, FDA Scrapbooks, Vol. 11.
[25] "A Better Drug Act or Only the Old One," *Am. Jnl. of Public Health* 26 (February 1936), 183-84.

Association with the bill.[26] There was even heavy criticism from *The Consumer,* an official publication of the Labor Department. The current bill was weaker than the existing statute, that journal charged, and should be condemned by the public. FDA's old ally, the *St. Louis Post-Dispatch* agreed.[27]

The only significant traditional dissenting voice not raised was that of Consumers' Research. This phenomenon was not a matter of a change in attitude by the group. CR simply had problems of their own for the moment. The fall of 1935 brought a permanent split between F. J. Schlink and his lieutenant, Arthur Kallet, over labor policy. CR was racked by an employees' strike which Schlink charged was Communist-motivated. Kallet took the side of the workers and their unionization drive. There was considerable violence on both sides. Eventually Kallet would go off to form a rival organization, Consumers Union, but in the meantime neither side had much time for additional opponents—a fact that must have cheered the hearts of the drug trade and, for that matter, FDA.[28]

The women's organizations, however, were united among themselves and in regard to their position on S. 5. They were pushing with an increased zeal. The tragic symbol of the new drive was the arrival in Washington of Mrs. J. W. Musser during the summer of 1935. Mrs. Musser had lost her sight from the use of Lash Lure. Eight subsequent operations had failed to help. She was in Washington to make the rounds of Congressional offices. Those to whom she spoke were properly impressed.[29] The women's organizations stepped up their efforts to dramatize the need for a new law. The American Association of University

[26] *Washington Daily News,* May 28, 1936, FDA Scrapbooks, Vol. 11.

[27] January 22, 1936, *ibid.,* Vol. 10.

[28] "Later Strike Developments," *General Bulletin of Consumers' Research* 11 (October 1935), 18; "Why Consumers' Faces Are Red," *PI* 173 (December 5, 1935), 90; "Consumers Union Reports," *PI* 175 (May 28, 1936), 64-65.

[29] *Wheeling* (W.Va.) *News,* June 2, 1935, FDA Scrapbooks, Vol. 7.

Women even put on a dramatization of the S. 5 House hearings at their South Atlantic Section convention.[30]

Other organizations such as the National League of Women Voters were flooding the country with brochures and bulletins in support of S. 5 and asking recipients to press the matter with their Congressmen.[31] The *Journal of Home Economics* cheered such efforts on and publicized new support for revision by bodies like Beta Lambda, a national organization of beauty culturists.[32] The periodical also warned its readers not to be misled into lessened effort because S. 5 had been watered down before Senate passage. That would help no one but the enemies of drug reform. The credo must be to hold what was left in the bill and to press with new vigor to correct the losses.[33]

The growing aggression of the women was complemented by the current state of the "guinea pig muckraking" literature. The literature itself had not changed, but it was a more aggressive force in the sense of cumulative effect from the ever increasing volume of such works. Certainly the scored industries were much more aware of the threat of that literature by 1936. As previously discussed, the prototypes for this material were Chase and Schlink's 1927 shocker, *Your Money's Worth* and Schlink's later volume, *100,000,000 Guinea Pigs.* The year 1934 brought such prominent additions to the list as M. C. Phillips' *Skin Deep, The Truth About Beauty Aids,* and James Rorty's *Our Master's Voice: Advertising. Printers' Ink* called Rorty's work a challenge for the industry to correct its own abuses. That periodical did not like "Comrade Rorty" but had to admit that many of the author's charges were valid.[34]

The volume was indeed a challenge. Rorty was not calling for a new law. He was talking about an end to competi-

[30] AAUW leaflet, *Consumers Beware, ibid.,* Vol. 10.
[31] National League of Women Voters pamphlet, *Needed: A New Food and Drug Law, ibid.,* Vol. 9.
[32] "Revision of the Federal Food and Drug Bill," *Jnl. of Home Economics* 28 (February 1936), 103.
[33] "Food, Drug, and Cosmetic Legislation," *ibid.,* March 1936, 176.
[34] "Comrade Rorty Lifts the Lid," *PI* 167 (May 10, 1934), 90.

tive capitalism. Left-wing intellectual Stuart Chase, reviewing the book, was delighted at the case which the author made. Advertising would stand or fall with the whole system of capitalism, Chase wrote. "Perhaps Mr. Rorty has really written an epitaph."[35] The words were enough to chill the heart of the most hardened opponent of legislative reform in the advertising field.

Before the dust had settled from the appearance of these works there were more on the way. In 1935 a triad of volumes emanating from the staff of Consumers' Research hit the market: *Eat, Drink and Be Wary* by F. J. Schlink, *Counterfeit—Not Your Money but What It Buys* by Arthur Kallet, and *Paying Through the Teeth* from the pen of Bissell Palmer. Reviews in *Printers' Ink* called Schlink's effort "prejudiced, inaccurate, and inexcusably sensational," and pronounced *Counterfeit* in its contents "aptly named."[36] Perhaps the criticism was warranted, but in terms of significance the matter was academic. The *New Republic* made the point: "The growing literature on consumer protection . . . is a symptom of mounting resentment against proprietary frauds."[37]

The guinea pig tide continued in 1936. Rachel Palmer and Sarah Greenberg published *Facts and Frauds in Woman's Hygiene*, while J. B. Matthews offered the public, *Guinea Pigs No More. Survey* had lavish praise for Matthews, whose volume "presents forcefully and bitterly the consumers' case against profit-making business."[38] Gleefully the *Nation* related that Hearst syndicate's *Drug World* in its hysteria over the Palmer-Greenberg volume had "rushed into print . . . the best unintentional publicity campaign the publishing industry has witnessed for some time." Two days before *Facts and Frauds* was published, *Drug World* even sent special delivery letters to all manufacturers

[35] "Advertising: An Autopsy," *Nation* 138 (May 16, 1934), 567-68.
[36] "Read and Be Weary," *PI* 173 (December 5, 1935), 67; "Counterfeit," *PI* 171 (June 6, 1935), 34.
[37] "The Consumer Pays and Pays," *New Republic* 84 (September 18, 1935), 166.
[38] "Consumers Unite," *Survey* 122 (July 1936), 221-22.

named in the book warning them of "the coming blow."
Alas, these efforts only helped to exhaust the first printing
immediately.[39]

These two volumes were merely prelude, however, to
the major literary explosion of 1936—Ruth Lamb's *American Chamber of Horrors*. While the work was not properly
a member of the immediate guinea pig family it was obviously a very close cousin. Just as obviously the popularity and impact of *Horrors* were rooted in the ground
swell of public interest created by several years of muckraking materials. Like the guinea pig authors, Miss Lamb
exposed many blatant abuses in the food, drug and cosmetic industries. More basic, however, was her exposure
of the whole history of the current efforts to pass a new
drug statute, telling consumers, according to *Business
Week*, exactly why they needed a new law and exactly why
they had not gotten it.[40]

The trade, particularly the drug industry, was enraged.
Drug and Cosmetic Industry set the tone; it took the book
as an FDA attempt to "whip" up public support for a
stronger bill. Manufacturers had been trying to cooperate
on a new law, the periodical moaned, and now this. "All
of which clearly indicates that the administration does not
give a damn for the industry or its cooperation."[41] But the
anger was not limited to the medicine manufacturers. Even
the dairy trade was upset over the "scurrilous attack" in
Horrors on the butter industry.[42] Not the least of those
unhappy was Ewin Davis of FTC. Miss Lamb had managed to take several swings at the activities of the Commission—statements which Davis considered "grossly and in-
tentionally inaccurate."[43]

The affected industries had reason to be upset. In the
light of the Lamb volume they not only suffered much pub-

[39] "Hearst and Lydia Pinkham," *Nation*, August 1, 1936, FDA
Scrapbooks, Vol. 12.
[40] "Chamber of Horrors," *Business Week*, February 22, 1936, 9-10.
[41] "Time to Fight," *D&C Industry* 38 (March 1936), 309.
[42] *St. Paul Dairy Record*, June 3, 1936, FDA Scrapbooks, Vol. 11.
[43] *Advertising and Selling*, May 7, 1936, *ibid*.

lic criticism but, in addition, the demands for a Congressional investigation of the food, drug, and cosmetic lobby were growing.[44] Moreover, the popularity of the Lamb volume was a promise that the stream of such exposure literature would not lessen in the future. It did not. In 1937 Peter Morell published *Poisons, Potions, and Profits*. Ruth Brindze gave the public *Not To Be Broadcast: The Truth about the Radio*, and Rachel Palmer, aided by Isidore Alpher, struck again with *40,000,000 Guinea Pig Children*. There was more of the same in 1938. The food, drug, and cosmetic trade was beginning to know the word "aggression" very, very well.

That the affected industries did understand the word and, even more vividly, the danger at hand doubtless explained why the sharpest divisions on a new law in 1936 were within the business community. More particularly the dialogue was in the proprietary medicine industry since the majority of the other concerned trades had joined the advocates of immediate law revision. Among the nostrum vendors, opposition to S. 5 was spearheaded by the United Medicine Manufacturers of America and the Institute of Medicine Manufacturers. In the late fall of 1935 these two organizations merged their resources to maximize their opposition efforts in the face of so much capitulation among their brother organizations. Their campaign was centered on halting attempts to strengthen multiple seizure provisions, on weakening the already watered down formula disclosure clauses of S. 5, and on eliminating FDA advertising controls as authorized in Copeland's measure.[45]

A drug journal poll showed that, in their opposition, the two organizations still spoke for 48 percent of the industry.[46] The percentage was new. The rhetoric of the opposition was old—by now threadbare. Even *Printers' Ink* noted editorially that "the position of the Institute of Medicine

[44] "Impure Food and Drug Laws," *New Republic* 87 (June 10, 1936), 137; "Chamber of Horrors," 9-10.
[45] *New York Times*, December 10, 1935; *Drug Trade News*, November 11, 1935, FDA Scrapbooks, Vol. 8.
[46] *Drug Trade News*, October 28, 1935, *ibid.*

Manufacturers has not changed one iota from the very beginning."[47] Clinton Robb, counsel for the U.M.M.A., sounded very much like an old record when he charged that the present bill aimed at the destruction of self-medication. His charges that the provisions of S. 5 were the written voice of the AMA were equally familiar.[48] Even the best of the opposition tactics were dated. When Mrs. Jenckes of the House introduced a new drug bill in August 1935, *Printers' Ink* was sure of the motive: "Somebody is gunning for the Copeland Bill and is going to try to kill it by fair means or foul." To that journal the effort was strictly bad news for the industry.[49]

Here was the basis of the trade division. Segments of the affected business community had long warned about what the future might bring if delay in passage of a new law continued. Now the future had come. Between January 1935 and October 1936, ninety-two laws touching the drug field were passed in thirty-nine states.[50] The great majority of these statutes were aimed at specific matters such as barbiturates, poisons, and the like, but the obvious next step was comprehensive legislation governing the whole drug field within the state. The *Oil, Paint and Drug Reporter* forecast that the trade could expect this very situation in the near future.[51] Indeed the "possibility" of comprehensive state and local statutes was no longer merely a threat. Louisiana was moving toward just such an act in the spring and summer of 1936. This bill was fashioned carefully on the current Copeland measure and designed to complement S. 5 if the Senator's opus became law. To *Drug and Cosmetic Industry*, the Louisiana situation demonstrated, among other things, the futility of continued opposition to S. 5. If the state bill became law, manufac-

---

[47] "For the Copeland Bill If—," *PI* 174 (January 2, 1936), 62, 64.
[48] "The U.M.M.A. Meeting," *D&C Industry* 37 (November 1935), 608; *House Hearings on S. 5* (1935), testimony by Robb.
[49] "And Now Mrs. Jenckes," *PI* 172 (August 1, 1935), 84-85.
[50] Sol Herzog, "New State Laws Affecting Pharmacy," *Am. Druggist* 94 (October 1936), 46-47.
[51] August 17, 1936, FDA Scrapbooks, Vol. 12.

turers would either have to keep their products and advertising out of Louisiana or to operate exactly as if the Copeland measure had passed into law.[52]

Nor was Louisiana the only available object lesson on the wages of further opposition. A new drug statute seemed imminent in New York City. By December 1935, the sole apparent reason that this "little Tugwell bill" had not become law was the decision of Mayor Fiorello La-Guardia to temporarily postpone action on the matter. LaGuardia was hardly an ally of the drug manufacturers. He made it plain that the postponement was simply to see what Congress would do on S. 5.[53] "It is regarded as a foregone conclusion," *Printers' Ink* warned, "that if the Copeland Bill does not make the grade, Mayor LaGuardia's measure will soon be law in New York City."[54] The *Oil, Paint and Drug Reporter* agreed.[55]

Those segments of the drug industry that saw this handwriting on the wall began to press for S. 5 with zeal. The new efforts of the nostrum vendors in the Empire State were strong enough to gain the interested notice of the *New York Times.*[56] These currents showed up strongly in the trade press. *Drug and Cosmetic Industry* as well as the *Oil, Paint and Drug Reporter* both urged the industry to get behind immediate passage of a federal law. The future was too dangerous.[57] *Drug Trade News* also warned that the lack of a federal statute was creating a witches' brew of "unfriendly" local legislation.[58] The Proprietary Association's James Hoge even wrote an article for *Printers' Ink* entitled, "How the Copeland Bill Extends Business

[52] "Copeland Act in Louisiana," *D&C Industry* 39 (July 1936), 88.
[53] *OP&D Reporter*, December 9, 1935, FDA Scrapbooks, Vol. 9; "S. 5 Passage Conceded," *PI* 173 (December 12, 1935), 12.
[54] "Mr. Rorty Has Fun," *ibid.*, November 28, 1935, 110.
[55] December 9, 1935, FDA Scrapbooks, Vol. 9; "S. 5 Passage Conceded," 12.
[56] November 10, 1935, 9.
[57] "The Drug Bill," *D&C Industry* 37 (July 1935), 29-30; *OP&D Reporter*, August 17, 1936, FDA Scrapbooks, Vol. 12.
[58] "Support the Copeland Bill," *PI* 173 (October 31, 1935), 98.

Opportunity"—a dramatic symbol that times had changed.[59]
Chances for a new law were better than ever before.
Not only was the opposition at its lowest ebb but consumer
support reached its highest pitch to that date. The single
immediate cloud was the advertising control issue. In this
matter FDA's trade opponents could achieve added force
by alliance with the Federal Trade Commission. Walter
Campbell argued the cause of his agency with eloquence
at the House hearings. He pointed out that FTC's advertising powers came not from a Congressional grant but
simply by implication out of the agency's authority to sup-
press unfair trade competition. He explained with care
how the Raladam court decision had greatly restricted the
ability of FTC to protect consumers from false advertising.
He elaborated on the need of FDA to possess regulatory
authority over advertising, several times emphasizing that
such authorization would take nothing from the Federal
Trade Commission.[60]

Judge Ewin Davis, Commissioner of FTC and a past
member of the House from Tennessee, was also at the
hearings. He was in a far more confident mood than Campbell. Davis defended the present operation of his organization in regard to regulation of advertising. The field was
covered, he noted, though if the Congress saw fit, some
"simple and clarifying amendments" to the FTC act might
be valuable. In a bow to the business community he af-
firmed his faith in the cease and desist procedure of his
agency, assuring the committee that he sought no criminal-
prosecution powers. The businessman was not after all a
criminal and should not be made to feel like one.[61] Finally,
he assured the committee, in rhetoric ringing less of hu-
mility than confidence about the future, that on the matter
of advertising control his agency would certainly abide by
the decisions of the Congress and the President. Davis did

[59] Hoge, *PI* 171 (June 13, 1935), 71-74.
[60] *House Hearings on S. 5* (1935), testimony by Campbell, 43,
75-78, 82-84.
[61] *Ibid.*, testimony by Davis, 646.

not feel this authority should pass to FDA *but*, "we want
to do what both of you want done."[62]

FTC had reason to be confident. It was not without significant trade support. The Commissioners had pressed and
would continue to press their cause on Roosevelt.[63] He had
not sided with Davis and company nor had he publicly
or officially rejected their cause. It was common knowledge
that the FTC had extensive backing in the House. As later
events would prove, the agency had important friends in
the Senate as well. Indeed one reason for Davis' confidence
at the House hearings may well have been that he knew that
in the new session bills would be introduced in both chambers to strengthen the authority of FTC.[64]

In the months following the House hearings various interested journals expressed their feeling on the probable
victory of the Trade Commission. By December, *Drug
Trade News* felt the only sure thing one could say of S. 5
was that the House would vote to place advertising with
FTC.[65] In mid-February 1936 the same periodical repeated
its prophecy.[66] During April, *Business Week* warned that
a "real fight" would be required to keep the advertising
provisions of S. 5 intact.[67] The *Journal of Home Economics*
echoed the same belief in May.[68] Throughout the spring
Standard Remedies increased its coverage on the activities
and operations of FTC.[69]

With Copeland's bill under wraps in the House committee and speculation on that measure's future growing, many

[62] *Ibid.*, 643.

[63] Summary of letter, FTC to Roosevelt, May 12, 1936, OF 375,
Roosevelt Papers.

[64] For a general record see Charles W. Dunn, *The Wheeler-Lea Act:
A Statement of Its Legislative Record.* Also see Lamb to E. M.
Kirkpatrick, March 1, 1938, Correspondence, Box 124.

[65] December 9, 1935, FDA Scrapbooks, Vol. 9.

[66] February 17, 1936, *ibid.*, Vol. 10.

[67] "Snagged," *Business Week*, April 4, 1936, 32.

[68] "S. 5," *Jnl. of Home Economics* 28 (May 1936), 319.

[69] Vol. 23, Spring 1936; see in particular "Digest of F.T.C. Stipula-
tions," April 1936, 13.

people felt the position of the President could be crucial.[70] The confusion was compounded, however, because virtually no one seemed to know what that position was. Even in retrospect it is difficult to say more than that FDR was mildly interested in the passage of *some* type law during 1936. This much was not totally clear at the time. As early as January 1936 some segments of the affected trades felt that passage was improbable because the Administration had lost interest in S. 5.[71] It would seem that House Speaker Joseph Byrns felt the same. In January he told the press the Copeland bill was not "urgent" legislation, and he doubted that it would pass in the present session. In response to questions on the future of advertising provisions Byrns further stated that he was "not exactly satisfied" with the bill as passed by the Senate.[72]

Robert Allen of the Pure Food League wrote FDR to make sure he was apprised of Byrns's comments.[73] Allen hoped, apparently, that he might spur the President to take some positive action. Charles Dunn sought to have Tugwell intercede with the Chief Executive as the one means of preserving advertising authority for FDA.[74] Senator Copeland also requested FDR's help in moving the bill through the House.[75] In the matter of getting S. 5 out of committee the President did take a hand. Committee chairman Sam Rayburn promptly responded to a February White House note, stating that machinery was already in operation.[76] In the matter of Presidential preference on the issue of advertising controls, however, no clear indication was given and confusion continued to reign.

Before the public, the Chief Executive refused to assume

[70] *Am. Perfumer*, January 1936, FDA Scrapbooks, Vol. 10; Dunn to Campbell, December 6, 1935, Correspondence, Box 565.
[71] *Am. Perfumer*, January 1936, FDA Scrapbooks, Vol. 10.
[72] *Food Field Reporter*, January 13, 1936, *ibid.*
[73] Allen to Roosevelt, January 21, 1936, OF 375, Roosevelt Papers.
[74] Dunn to Campbell, December 6, 1935, Correspondence, Box 565.
[75] Dunn to Campbell, February 6, 1936, *ibid.*, Box 711.
[76] Roosevelt to Rayburn, undated, with notations by Marvin McIntyre dated February 7, 1936, OF 375, Roosevelt Papers.

any responsibility in the future of S. 5. The trade journals picked up his backstage intervention with Rayburn, but questioned by the press in March, FDR professed to know nothing about the status of the bill.[77] The flippancy of his comments suggested a lack of interest in the whole matter. Again questioned in May on whether a bill would pass at the present session the President contented himself with the brief comment that, "Hope springs eternal. I have been hoping for three long years."[78] By June a trade rumor had it that Roosevelt was not, in fact, interested in having S. 5 pass. If the bill became law it would carry Copeland's name, and Copeland was *persona non grata* in the White House for political reasons. *Printers' Ink* did not believe the President would be that petty. It was true, however, that by June, Copeland had openly bolted from the New Deal. He had even refused to attend the Democratic party convention, indicating his preference to attend a medical meeting in Cleveland scheduled the same week.[79]

Whatever the case here, there was only minimal White House direction in the handling of S. 5 during 1936. The real question insofar as *Drug and Cosmetic Industry* was concerned was who, if anyone, had provided the leadership for the measure as it came from the House committee. The bill had been totally rewritten. Obviously the drug industry had exercised little voice in the matter. The new variation clause would allow official standard variations in strength but not in quality and purity as permitted in the Senate bill. The multiple seizure clause was strengthened also. The Senate version allowed such seizure only where an article was "imminently dangerous to health." The House substitute added the words "or in a material respect, false, misleading or fraudulent."[80]

[77] *Franklin Roosevelt Press Conferences*, March 10, 1936, Vol. 7, 185, Roosevelt Papers; "Copeland Bill Revived," *PI* 175 (April 23, 1936), 12; "S. 5 Wrangle," *Am. Druggist* 93 (May 1936), 27.
[78] *Franklin Roosevelt Press Conferences*, May 1, 1936, Vol. 7, 230, Roosevelt Papers.
[79] "It Looks Bad for S. 5," *PI* 175 (June 18, 1936), 12.
[80] "The Legislative Scramble," *D&C Industry* 38 (June 1936), 760-61.

The food and advertising industries had not with certainty had their way. Unlike the Senate measure the House bill provided the Secretary with power to establish multiple quality grades for food products. Organized medicine had not held sway, or the provision to allow proprietary manufacturers to file their formulas with the Secretary in lieu of label disclosure would not have remained. Finally, and quite apparently, the House version was not the voice of FDA. That agency had lost the power to regulate advertising as proposed in Copeland's version.[81] In this last assumption at least *Drug and Cosmetic Industry* was quite correct. When Walter Campbell saw the new bill he was far from happy. "In my judgement," he wrote Secretary of Agriculture Wallace in May, "every effort should be made to kill the measure."[82] What *Drug and Cosmetic Industry* had overlooked, however, was that the committee did allow one victor—the Federal Trade Commission.

The moving force behind the House version does not seem, however, as difficult to locate as *D&C Industry* made out. Virgil Chapman had been deeply impressed with the need for a new strong food bill during the House hearings in the summer of 1935. It seems logical to assume that as chairman of the House subcommittee he was responsible for the strengthened provisions of the lower chamber version. But Chapman was also a realistic politician. In omitting the advertising provisions he was simply bowing to the overwhelming sentiment of the House, and perhaps the full Commerce Committee. S. 5 could never pass the lower chamber with the advertising provisions included. The Kentucky Congressman would dissent against the omission of these controls, but in truth it would appear that he had chosen to get passed what he could and pin any real hope for restoration of FDA authority over advertising on the House-Senate conference committee.

The House measure was reported from committee on May 22, 1936, along with a minority dissent on the adver-

[81] *Ibid.*
[82] Campbell to Wallace, May 25, 1936, Commissioners' File, Box 12.

tising section by Virgil Chapman, Carl Mapes, and Schuyler Merritt. The majority endorsement did not necessarily mean that passage was assured. The immediate problem was to get the bill to the floor. This was unlikely unless Roosevelt intervened. One obstacle was the unwillingness of Speaker Byrns and Representative John O'Connor of the House Ways and Means Committee to bring the measure up out of sequence without a request from the President. Both Representative Rayburn of the Commerce Committee and Senator Copeland besought the Chief Executive in early June to lend this assistance, or, as Copeland put it, the bill "is lost."[83] The Senator was not happy over the House substitution on advertising, but he believed an agreement could be reached between Senate and House conferees.[84]

FDR assured Copeland on June 16 that he was "doing everything possible to get action."[85] Indeed, it appears that the intercession of the President was solely responsible in bringing S. 5 to a vote. There was a profound irony here. For the first time since 1933 many proponents of a new bill did not want intervention. Certainly the AMA and most consumer groups such as Consumers Union preferred that S. 5 not pass into law. *American Professional Pharmacist* labeled the bill "an impotent tool, nebulous in terminology, and valueless for any effective prosecution."[86] CU called it a "legislative monstrosity."[87] Even *Christian Century* stated that passage of the House version would be "a surrender to the most unworthy elements in the world of food and drug manufacturing."[88] Nor did FDA like the measure as

[83] Rayburn to Roosevelt, June 3, 1936; also Copeland to Roosevelt, June 3, 1936, both in OF 375, Roosevelt Papers.

[84] Copeland to Roosevelt, June 9, 1936, *ibid.*

[85] Roosevelt to Copeland, June 16, 1936, Petitions and Memorials File.

[86] "R.I.P., S-5," *Am. Professional Pharmacist* 2 (June 1936), 13.

[87] Otis Pease, *Responsibilities of American Advertising*, 123; Lamb to Mary Ross, June 12, 1936, Correspondence, Box 716; Consumers Union *Reports*, June 19, 1936, FDA Scrapbooks, Vol. 11.

[88] "Is There to Be a New Food and Drug Law," *Christian Century* 53 (June 10, 1936), 833-34.

it stood. Campbell hoped S. 5 would perish in the House because he believed a stronger bill could be passed in 1937.[89]

For Campbell, the AMA, and the consumer groups the worst aspect of the House measure was that advertising had been left to the Federal Trade Commission. That agency's cease and desist procedure would not provide the consumer with adequate protection. Oddly enough, some manufacturers and publishers were unhappy with S. 5, and thus Presidential intervention, for opposite reasoning. They felt FTC's increased power would make the agency a tyrant over the whole field of business. They much preferred that advertising authority rest with FDA.[90] Roosevelt also preferred the Food and Drug Administration, but in his intercession made no effort to press that preference on the House. He sought merely to bring S. 5 to the floor.[91]

On Friday, June 19, the House Speaker received word from the President that he would like to see action.[92] That same afternoon the Copeland bill was brought to the floor under a suspension of chamber rules. Debate was limited to forty minutes, and the rule suspension procedure prevented amendments. At the fall of the gavel the measure was passed by an overwhelming majority. The Senate was notified and asked to concur in the House amendments. As expected, the upper chamber disagreed and asked for a conference. On Saturday, Senate and House conferees met. That night Copeland announced to his colleagues that agreement had been reached on all points save one, the advertising provisions. On this item, he continued, the conferees had "worked out what on our side of the table we considered to be a very happy compromise."[93]

The agreement provided that all advertising matter rela-

[89] Campbell to Wallace, May 25, 1936, Commissioners' File, Box 12.
[90] Pease, 123.
[91] J. J. Durrett to Harvey Cushing, June 24, 1936, Correspondence, Box 713.
[92] Lamb to Crump Smith, July 7, 1936, *ibid.*, Box 716.
[93] *Congressional Record*, 74th Cong., 2nd Sess. (June 20, 1936), 10,514.

tive to health would be regulated by the Food and Drug Administration. All matters having to do with economic problems as well as material relating to food and cosmetics would be administered by the Federal Trade Commission. Copeland moved that the Senate accept the conference report. The motion carried. The real drama came in the House. There, attack on the compromise was led by Representatives Reece and McReynolds of Tennessee, home state of FTC's Ewin Davis as well as a center of proprietary medicine manufacturing. Virgil Chapman tried desperately to defend the agreement of the conferees, but by the end of the debate quite obviously he had lost. Representative McReynolds sounded the keynote for the opposition. "Now, members of the House," he cried, "what are you going to do about it? Are you going to turn this over to Tugwell for enforcement or are you going to leave it with the Federal Trade Commission with such men as Judge Davis and other men from this House on that Commission?" The 190 to 70 division was a thundering endorsement of McReynolds' logic.[94] S. 5 was dead.

At the end all that remained was post mortem. Who killed the Copeland bill? One interesting aspect of the autopsy was the indignation of a sizable portion of the affected industries. This indignation came from that segment which had only recently come to believe in the necessity of a new federal law. They had joined the S. 5 camp out of fear about the future—fear of state legislation and fear that next year an even stronger bill might be introduced in the Congress. They were angry about the current turn of events. The *Oil, Paint and Drug Reporter* spoke for them when it denied any industry responsibility for the legislative collapse: "The blame and the shame for the failure falls wholly upon Congress, which permitted the influence of rival governmental agencies to outweigh its regard for the paramount public welfare."[95]

[94] *Ibid.*, 10,679-80.
[95] June 29, 1936, FDA Scrapbooks, Vol. 11.

Printers' Ink laid particular blame on FDA. That agency had accepted S. 5 in the Senate, then turned against it in the House and ultimately destroyed any possibility of passage by allowing publication of Miss Lamb's book. *American Chamber of Horrors* aggravated the dislike and distrust of House members for the Department of Agriculture as well as antagonizing the Federal Trade Commission. In conclusion *PI* agreed with the *Oil, Paint and Drug Reporter*.[96] Agency squabbling and not the trade had provided "the rock upon which the bill at last came to grief." Even several nontrade periodicals such as *Christian Century* scored the interagency antagonism. After resisting onslaughts by the advertising industry and trade manufacturers S. 5 "finally fell a victim of bureaucratic jealousy within the government itself."[97]

In a sense these critics were correct. Whether justified in the light of consumer protection or not, the respective power quests by FDA and FTC had played a major role in the defeat of S. 5. Campbell's agency was apathetic about the bill's advertising clauses though the Federal Trade Commission must bear even more of the blame. FTC was unwilling, as Ruth Lamb wrote a confidant, "to accept anything less than complete control, even though it meant the defeat of a measure so vital to life and health."[98] Representative Sam Rayburn of Texas, Chairman of the House Commerce Committee, saw the matter clearly. "There might be a little lobbying around here by some people," he told colleagues, "but there is nobody who has lobbied around this capitol on any bill in the 23 years I have been in Congress more than members of the Federal Trade Commission have lobbied on this bill, and I love the Federal Trade Commission."[99]

The failure of S. 5, however, involved more than mere differences between government agencies. Contrary to the

[96] "A Word to the Wise," *PI* 175 (June 25, 1936), 98-99.

[97] "Who Killed the Copeland Bill," *Christian Century* 53 (August 12, 1936), 1079-80.

[98] Lamb to Crump Smith, July 7, 1936, Correspondence, Box 716.

[99] *Congressional Record*, 74th Cong., 2nd Sess., 10,680.

claims of the trade press, continued opposition to the bill, particularly within the proprietary medicine industry, did play a part in the defeat. Certainly the failure of FDR to provide Congressional leadership on the advertising issue, even though he favored the food and drug unit, was a major factor.[100] In the end ultimate responsibility rested with the personal prejudices and adamancy of the House membership. In this regard perhaps Royal Copeland summed up the matter best: "Tugwell and Tennessee beat us."[101]

[100] Durrett to Cushing, June 24, 1936, Correspondence, Box 713.
[101] Crawford to Cavers, June 25, 1936, Commissioners' File, Box 12.

VI

"MUCH POWER TO YOUR ELBOW"

I judge from all accounts that the Federal Trade
Commission are as great obstructionists to the public
welfare as the majority of the Supreme Court appear
to be. To this all I can say is, damn the lawyers.
Much power to your elbow.

DR. HARVEY CUSHING TO
J. J. DURRETT OF FDA,
JUNE 24, 1936

"WE ARE attempting to prepare exhibits of some . . . preparations for which extravagant therapeutic claims are made
in collateral advertising," FDA's J. J. Durrett wrote the
chief of the drug unit's Central District in the fall of 1936.[1]
The letter was one of many which passed from the Washington office to field stations over the country. Field offices
were asked to forward printed materials on twenty-eight
categories of food, drug, and cosmetics but, as Durrett put
it, "we are particularly anxious to obtain newspaper or
periodical advertising that is being used in furthering their
sales." Area stations were to give this task "all reasonable
priority" since Walter Campbell hoped to have the exhibit
complete by January 1, 1937—when presumably a new
food-drug bill would be introduced into Congress.[2] A little
more "elbow power" could be a valuable thing.

So it was that in the fall of 1936 the Washington office
of the Food and Drug Administration began to take on the
aura of a giant medicine show. Labels, pamphlets, and
press advertising arrived with every mail. From Atlanta

[1] Durrett to Central District Chief, November 10, 1936, Correspondence, Box 715.
[2] Charles Crawford to Chiefs all Districts, November 11, 1936,
ibid.; Durrett to Central District Chief, November 10, 1936, *ibid.*

[126]

came Blakeley's Acid Iron Material as well as 4-44, a panacea for stomach sufferers. According to advertising "4-44 revitalizes your food with sixteen minerals without which you cannot have a sound stomach or vigorous body." Its label, however, bore no therapeutic claims.[3] Perhaps the sufferer's problem was simply a common cold. If so, there was always the proprietary, Frog Pond, whose Baltimore advertising copy touted the balm as a "sure cure for Chill–Fever–Cold" and even prickly heat or rheumatism.[4] Rheumatism as well as neuritis and arthritis could also be alleviated by Casey's Compound. Like 4-44, Casey's listed no therapeutic claims on its packaging. "Sufferers" could learn of the wonder-working possibilities of this medicine only by writing for a "free pamphlet" from the manufacturer.[5]

If one's affliction was in the intestinal tract, relief was also simply a drugstore away. Periodical advertising from the F. E. Young Company of Chicago announced that the public could purchase Dr. Young's Rectal Dilators on a money-back guarantee. Through "natural methods" the dilators assertedly strengthened rectal muscles by "imitating Nature's own process." The result was to promote regular bowel habits and thereby end suffering from constipation or piles.[6] As an alternative to the dilators the afflicted could purchase, according to copy in Atlanta, Triple Strength Chinese Herb Compound. The compound was beneficial for stomach troubles in general—indigestion, dysentery, biliousness, dyspepsia—but equally useful for gallstones, malaria, asthma, eczema, rheumatism, and bad breath.[7] About the only problem which this proprietary could not handle was bad skin. In which case the unhappy could turn to Gouraud's Oriental Face Cream. The company's advertising claimed it to be an invaluable aid in promoting beauty. The AMA's Bureau of Investigation claimed that

[3] J. J. McManus to Chief, FDA, Washington, November 5, 1936, *ibid.*
[4] Baltimore Station to Washington Office, October 29, 1936, *ibid.*
[5] E. O. Eaton to Chief, Western District, December 16, 1936, *ibid.*
[6] Advertisement for Dr. Young's Rectal Dilators, *ibid.*
[7] McManus to Washington Office, November 17, 1936, *ibid.*

Gouraud's balm more properly promoted discoloration of the skin as a result of the mercury in its composition.[8]

The Washington offices of FDA could offer further a variety of medicinal remedies for the sexually inadequate. There were, for example, the advertising pamphlets of the Sherwood-Arthur Pharmacal Company, makers of Persenico. With its hemoglobin base Persenico had proved highly satisfactory, the makers declared, "in combating neurasthenic impotence, pre-senility, low vitality and general nervous ailments, particularly . . . of sexual origin." The manufacturer insisted that in most cases results would "begin to be felt" within ten days.[9] A better choice, perhaps, was Revivio, a love-sex hormone "to improve your vigor," by Dr. Siegal's Medical Products. Revivio made no claims of ten-day results. Change would take a while. In the meantime, however, the sexually troubled male could gain immediate relief through the temporary expedient of Dr. Siegal's "Great Mechanical Developer," an added attraction for immediate purchase of Revivio. As advertising explained, the impotent male inserted the penis into the developer and turned the pump. The vacuum drew more blood to the organ and brought about an erection.[10] So the list of panaceas grew in Washington, collectively announcing the end of human infirmity and the obsolescence of the medical profession.

Out of such bizarre, though commonplace, proprietary publicity FDA created a new exhibit designed to illustrate shortcomings in the existing drug statute. It might have been called Walter Campbell's Compound for Increasing the Vitality of His Agency. Hopefully the display would "promote" a new drug law which included effective advertising controls. The material was primarily prepared for presentation to Congressional committees. Certainly this exhibit was never publicized like the 1933 "Chamber of Horrors." This fact perhaps explains why it attracted vir-

[8] B. O. Halling to A. W. Garrett, December 11, 1936, *ibid.*
[9] Pamphlet on Persenico, *ibid.*
[10] A. W. Garrett to Washington Office, February 10, 1937, *ibid.*

tually no attention in the trade press. FDA had learned the necessity of discretion. Yet the exhibit in photographic form was made available to the public. By January 1937, agency personnel were quietly but actively soliciting "customers" among private organizations.[11]

An additional reason that these activities by the drug unit drew minimal attention in the concerned industries was the changed posture of much of the trade. Large numbers of manufacturers were far more irritated that a new bill had not passed already than at the prospects of a bill passing in the near future. The posture change since 1933 was well illustrated in a November *Printers' Ink* editorial that announced Copeland's expectation of early enactment for a new food bill he would introduce in January. *PI* declared, "We hope so." The logic of the editorial was the threat of state legislation if no federal law passed. Such legislation differing as it would from state to state would be a disaster.[12] The fear echoed by *PI* was not brand new in 1937 but there was a new ground for it. Consumer groups, particularly the women's organizations, were adding force to their "elbows" in a new line of attack. By way of flank assault the women planned to introduce model food and drug bills in every state legislature which met in January.[13]

The direct effort for strong federal legislation, of course, continued. Here proponents of reform were more vocal, as well as more unified, than ever. Now steadfast allies in the profession of pharmacy had put the Drug Trade Conference on record during 1935 as in full support of the then current bill, S. 5. Leading representatives of the profession, including the APhA, the National Association of Boards of Pharmacy, and the American Association of Colleges of Pharmacy, reaffirmed that position in 1936. By early 1937 the *American Journal of Pharmaceutical Education* was

[11] Theodore Klumpp to Dr. John Peters, September 25, 1936, *ibid.*, Box 862.

[12] "Let's Have It," *PI* 177 (November 12, 1936), 126.

[13] *Philadelphia Record*, December 27, 1936, FDA Scrapbooks, Vol. 13.

urging its readers to make an aggressive fight for the new Copeland bill. A position of positive support, the *Journal* warned, was the only acceptable position for pharmaceutical educators who realized that pharmacy was an "intimate health service profession."[14]

The *Journal of the American Pharmaceutical Association* also loudly demanded prompt passage of a bill similar to the Senate bill of 1935. Further delay in enactment of a new statute would simply represent a "disregard of public welfare."[15] Writing in the *American Journal of Pharmacy* in early 1937, Charles LaWall clearly exemplified the irritation of professional pharmacy with those who continued to oppose a new law. He was reacting to a recent magazine article which charged that the New Deal intended to make the public into guinea pigs for experimental purposes. "The writer overlooked the fact," LaWall snorted, "that 100,000,-000 of us are already guinea pigs living in the pens of the quacks and fakers."[16]

If the women's organizations and the agents of professional pharmacy were more aggressive in 1937, the militant Consumers' Research seemed to be mellowing. At least *Printers' Ink* thought so. Schlink's new magazine, *Consumers' Digest*, had much of the "usual malarkey about advertising, arsenic and saving your anti-freeze," but the militant leftist orientation of its publisher had decidedly diminished. Schlink no longer quarreled with advertising per se. He admitted that nostrums were worse in Russia than in the United States. He "casually lumps communism, fascism and consumer cooperatives together." "It seems just barely possible," PI concluded, "that Mr. Schlink wants someone to love him."[17]

The change was perhaps overdramatized by *Printers' Ink*, but some change there was. In the late fall of 1936,

[14] Editorial, *AJPhE* 1 (January 1937), 97-98.
[15] "Pure Food and Drug Legislation," *JPhA* 26 (February 1937), 105.
[16] LaWall, "Fads and Frauds in Foods and Drugs," *AJP* 109 (March 1937), 123.
[17] "Mr. Schlink's New Paper," *PI* 178 (January 14, 1937), 49, 50.

Consumers' Research even had some kind words for FDA: "That agency comes nearer to having a mandate to function in the consumers' interest than any other departments of the government." The occasion for the comment was that CR had endorsed the placement of advertising control authority with the drug unit. Consumers' Research backed also the drive of the women to introduce drug bills into state legislatures.[18] Perhaps the leftist strike at CR was responsible for Schlink's mellowing. Whatever the reason for the change, it did strengthen the forces pushing for a new food-drug bill. Trade opponents had a right to be apprehensive about the growing-strength of their antagonists, and they were. More recently, enemies of revision had become concerned that their consumer adversaries would use their growing strength in an attempt to shift the whole matter of food, drug, and cosmetic control out of the Agriculture Department and into the Public Health Service. There, bemoaned the *Chicago Journal of Commerce,* control would be much more subject to the influence and desires of the ever dreadful American Medical Association.[19]

Trade fears of such a plan had no basis in fact and were totally unrealistic. Militant consumer power was stronger than ever, but not that strong even if the desire was there. The hard fact was that the force of consumers never reached a scope sufficient to put through any bill without support in the affected industries. Any effort at a shift to the Public Health Service would have been abortive from the start. Current trade support would have been lost and that support was an all important factor. That the opposition to statutory revision had greatly overestimated the power of the reformers was well illustrated by current thoughts on strategy within the Food and Drug Administration. Though the matter was never publicized, it would appear that Walter Campbell had gone full circle in his approach toward food-drug bill reform. He was coming

[18] "Off the Editor's Chest," *General Bulletin of Consumers' Research* 3, n.s. (October 1936), 2.
[19] October 5, 1936, FDA Scrapbooks, Vol. 12.

to believe that amendment to the 1906 statute might be a better strategy than continuing the fight for a completely new statute. He expressed this thought to Mrs. Harvey Wiley in September 1936.

Mrs. Wiley had written Ruth Lamb about a recent request from Copeland for suggestions in drafting a new bill. Campbell chose to answer personally and to inform her that he believed it was now possible to make do with an amendment. On the basis of their past three years of experience, he stated, "the highly skilled legislative draftsmen of the House and Senate may be able to formulate a satisfactory bill by amending rather than rewriting the law." FDA would not object to this procedure. "Furthermore," the drug chief continued, "there would be . . . a distinct legislative advantage in such a course, as it would cut the ground from under the arguments of some of the patent medicine crowd who have held that amendments . . . would be ample."[20]

In brief, Walter Campbell was still deeply concerned about the strength of trade opponents. FDA had taken a beating on the multiple seizure sections of S. 5 and any attempt to reverse significantly the Bailey-Clark compromise of 1935 would be at best an extremely difficult fight. If an amendment to the 1906 statute were introduced without direct reference to multiple seizure, FDA would stand a better chance of retaining unrestricted multiple seizure powers as established in the Wiley law. At least this way a certain weight of precedent and usage would be on the side of the food and drug unit. A second consideration was the matter of controls over food, drug, and cosmetic advertising. Here, too, the amendment approach seemed valuable. FDA might better be able to rally trade support, having gone to the long-demanded amendment form of revision.[21]

Perhaps another major factor that bore on Campbell's

[20] Campbell to Wiley, September 23, 1936, Correspondence, Box 714.
[21] "Drug Bill Friends Shift Tactics," *Business Week*, November 28, 1936, 19-20.

thought in the fall of 1936 was that Rexford Tugwell had
resigned from the government. He was joining ex-brain
trusters Charles Taussig and Adolph Berle in the American
Molasses Company.[22] *Business Week* saw this as an advantage for FDA since the Tugwell tag would be removed
from revision efforts.[23] *Food Industries* felt also that the
resignation would facilitate passage of a new bill.[24] Yet
surely for Campbell the exit of Tugwell was at best a mixed
blessing. The now Under Secretary was controversial, but
he had also been the spark plug of food-drug bill revision.[25]
Food Industries plainly admitted that the resignation meant
a less rigorous—though their choice of words was "more
reasonable"—bill.[26]

The worst aspect of Tugwell's departure, as Walter
Campbell must have known, was that "Rex" represented
the main line of communication between proponents of a
new law and President Roosevelt. The reformers needed
Roosevelt and he was at best "cool" to revision efforts.
Tugwell had kept the matter before the President. Without Tugwell, any aid from the White House was questionable. Certainly Royal Copeland could not fill the communication void. He was already politically distasteful to the
President and would become more so in 1937. Walter
Campbell must have been aware of the great loss involved,
as must many other drug reform proponents. It seems more
than coincidence that Mrs. LaRue Brown of the National
League of Women Voters wrote Miss Marguerite LeHand
of the White House staff in December asking for "advice"
on some new approach to the President. The League had
been unable to establish any line of communication with
FDR in the past, and Mrs. Brown felt that in regard to the

[22] "He's in Molasses Now," *PI* 178 (January 7, 1937), 105.
[23] "Drug Bill Friends Shift Tactics," 19-20.
[24] "Tugwell Joins Industry," *Food Industries* 8 (December 1936),
644.
[25] Paul Hutchinson to T. Swann Harding, October 23, 1936, Harding Papers.
[26] "Tugwell Joins Industry," 644.

food-drug bill some personal and informal relationship with the President was important.[27]

Whatever the rationale and factors involved in the consideration of an amendment effort, it never got off the ground. Such a move did have some support among proponents of tighter control outside FDA. The champion of the drug unit's cause in the House, Virgil Chapman, was converted to such a tack at least by the summer of 1937. In the light of the continued successful FTC lobbying in his chamber, Chapman introduced an amendment bill designated H.R. 7913.[28] Royal Copeland, however, was adamant in his opposition to such a proposition and remained so. When E. Fullerton Cook of the USP revision committee sent Copeland a copy of H.R. 7913 for his comments in July the Senator's reaction was swift and to the point. "I am unutterably opposed to H.R. 7913," he wrote Cook. "It is along the line of the old Mead Bill and is being used by the patent-medicine groups to defeat the legitimate bill. I hope you will not fall into the trap and accept this proposal."[29] Copeland knew that Chapman was not a tool of the medicine interests. Rather, the New York legislator was determined to put through a complete new bill—his bill. It seems plausible to assume that on this rock, whatever other factors involved, the amendment approach was doomed.

Copeland had resumed his efforts for a new statute in the early fall of 1936. While FDA was soliciting material for a new exhibit from field stations, the New York Senator was soliciting trade views in the drafting of a new bill.[30] Again the drafting process was to be carried out, according to Copeland's office, independently of the Food and Drug Administration. *Drug and Cosmetic Industry* applauded

[27] Brown to LeHand, December 11, 1936, OF 375, Roosevelt Papers.
[28] National League of Women Voters *News Letter*, FDA Scrapbooks, Vol. 16.
[29] Cook to Copeland, July 28, 1937, Petitions and Memorials File, 75A-F6.
[30] "To Senator Copeland," *PI* 177 (September 24, 1936), 15-16; *Drug World*, October 23, 1936, FDA Scrapbooks, Vol. 12.

the idea. USDA as a whole should be kept out of the picture, though the periodical was doubtful that the drafting would be completely divorced from the department.[31] In fact, Copeland did just as he promised. FDA received the full text of the new measure only after it was introduced in Congress.[32] The final writing of the Senate measure, as well as the drafting of a companion bill to be introduced in the House by Virgil Chapman, was done not in the Agriculture Department but by the Senate and House legislative counsels.[33]

The Copeland measure was ready by December and placed in the hopper on the fifteenth of that month.[34] The upper house version was again, and confusingly, designated as Senate bill 5. Chapman's measure was numbered H.R. 300. Reaction of those concerned with food and drug matters centered immediately on three points. First was the advertising provisions. Here there was a difference between S. 5 and H.R. 300. Both placed regulatory powers with FDA, but Copeland's bill stipulated that control would be handled by injunction while Chapman's version provided for civil and criminal judicial penalties. The second point of interest was seizure provisions. H.R. 300 allowed multiple seizure where goods were deemed "imminently dangerous to health" and so perpetuated the old Bailey-Copeland compromise. In S. 5 there had been a modification. The word "imminently" had been dropped from the phrase. The third point of concern was the variation clause in both bills which allowed strength variations from official standards. Proprietary goods were not subject to adulteration charges so long as strength conformed to any standard printed on the label. For both foods and drugs, however, full disclosure of formulas was demanded.[35]

[31] "New and Non-Official," *D&C Industry* 39 (October 1936), 438.
[32] Ruth Lamb to Ruth Lerrigo, January 16, 1937, Correspondence, Box 863.
[33] "New and Non-Official," 438.
[34] "Ready for the Gun," *Am. Druggist* 94 (December 1936), 27-28.
[35] Ole Salthe to Copeland, Petitions and Memorials File, 75A-F6; "Fourth Food and Drug Fight," *Business Week*, January 16, 1937,

Trade reaction was strongest on the matter of advertising. Here, as before, the dialogue was mainly within the drug industry. Both the food and the advertising men generally favored control in the hands of FDA and thus had no particular grievance with the new bills. Indeed in March 1937, when Mrs. Jenckes reintroduced her old drug measure in the House, Charles Dunn, who had originally drafted that measure, wrote Copeland immediately to deny any part in the current situation.[36] The drug industry was deeply and heatedly split on location of advertising control authority. Standing solidly with FDA was the American Drug Manufacturers Association, and the Toilet Goods Association. Officially the Proprietary Association sided also with FDA. That body requested its members not to seek changes which would endanger passage of the bills, but there was division within the membership in regard to advertising.[37] Aside from some members of the Proprietary Association, the support for the FTC largely came from two groups. They were the National Wholesale Druggists Association and the Institute of Medicine Manufacturers.[38]

The traditional proponents of food-drug law revision were pleased with the location of advertising powers in FDA. One exception was Arthur Kallet's militant Consumers Union which insisted that the whole matter of food, drug, and cosmetics regulation should be lodged in a Consumer Bureau inside the Public Health Service. Under the auspices of CU, Representative John Coffee introduced a

38ff.; "The Food and Drug Bills," *PI* 176 (January 14, 1937), 12ff.; "New Drug Bill," *D&C Industry* 40 (January 1937), 96, 97, 99; *Labor*, February 23, 1937, FDA Scrapbooks, Vol. 14.

[36] Dunn to Copeland, March 11, 1937, Correspondence, Box 711.

[37] "Reviving the Drug Act," *D&C Industry* 39 (November 1936), 583; "Proprietary Bulletin," *PI* 176 (January 14, 1937), 21; "Statement on the New Food and Drug Bills," *Standard Remedies* 23 (January 1937), 1.

[38] *New York Journal of Commerce*, October 5, 1936, FDA Scrapbooks, Vol. 12; "F.D.A. vs F.T.C.," *D&C Industry* 39 (February 1937), 189-90; Lamb to Charles Krewson, January 4, 1937, Correspondence, Box 863.

bill in the House to achieve that end during March.[39] This point of view was generally limited to Kallet's group and attracted little support otherwise. Consumers Union was unhappy also about the injunction procedure prescribed in S. 5.[40] Here the organization had much more support. Even Royal Copeland was defensive about this matter and justified the change solely on the basis that it would increase the chances of getting his bill passed.[41]

By the end of 1937, the American Medical Association would concede that S. 5-1937 was a better bill than S. 5-1936 but in the early months that organization found much room to criticize Copeland's latest opus.[42] The AMA was not pleased with the injunction process but their greatest grievance pertained to the variation clause. It was totally and completely inadequate. Indeed it was dangerous. The physician prescribed drugs, *JAMA* complained, on the basis of his knowledge of the United States Pharmacopeia. The variation clause meant that doctors would have no idea of the drug potency which their patients actually received.[43] The posture of the AMA as usual angered FDA. Charles Crawford expressed his feeling in a letter to Copeland's legislative assistant, Ole Salthe, during February: "This crowd, like so many of the industrial groups, is more concerned with sniping at minor issues than at getting a law with sound fundamentals which could be improved by future amendment."[44]

The irritation of the AMA with Copeland's legislative efforts was not new and thus represented no real loss for the revision camp. A more significant problem was the mo-

[39] "Arthur Kallet: His Bill," *PI* 178 (March 11, 1937), 80.
[40] Consumers Union *Reports*, March 1937, FDA Scrapbooks, Vol. 15.
[41] Copeland to Mrs. F. W. Fritchey, January 26, 1937, Petitions and Memorials File, 75A-F6.
[42] "Report of Officers," *JAMA* 108 (May 1, 1937), 1517; "Proceedings of the Atlantic City Session," *ibid.*, June 19, 1937, 2144.
[43] "National Food, Drug, and Cosmetic Legislation," *ibid.*, February 6, 1937, 476-78.
[44] Crawford to Salthe, February 8, 1937, Commissioners' File, Box 12.

mentary anger of the women's organizations, Copeland's strongest friends and allies. In their eyes the New York Senator had simply gone too far in his compromises, though their anger was not directed at him personally. The women were opposed to the injunction procedure on advertising control. It was too weak, they argued, in that the method halted a given advertising practice but not the sale of the product. They also objected to the variation clause for essentially the same reasons advanced by the AMA.[45]

Their greatest grievance was with the seizure provisions of S. 5. Granted the present clause was a bit better than the Copeland-Bailey compromise, but what the women wanted was unrestricted multiple seizure. The 1906 statute allowed this, and there should be no change. Copeland professed to want a stronger law himself but insisted S. 5 was the best he could presently hope to pass.[46] The zealous demands of the women were more than he would accept. Herein was the basis of an acute irony in February. Copeland was again attempting to halt public hearings on his bill but, for the first time, his antagonists were the women's organizations and not the trade.[47]

The alienation of the women was important, but it was not crucial. These organizations wanted a stronger bill but they were realistic. They were not likely to withdraw permanently their support from Copeland. To do so might well mean that an even weaker bill would become law or perhaps that no bill would pass. At least compromise was possible between Copeland and the women. They would soon be once more behind S. 5 even though the Senator might have had some qualms about the way they settled some of their grievances about the bill. The basis of Copeland's real sorrows in 1937 had little to do with his friends.

[45] "For an Amended S. 5," *PI* 178 (April 19, 1937), 12; "Food, Drug, and Cosmetic Bills," *Jnl. of Home Economics* 29 (April 1937), 250-51.

[46] *Congressional Record*, 75th Cong., 1st Sess. (March 9, 1937), 2020-21.

[47] Louise Baldwin to Copeland, February 10, 1937, Petitions and Memorials File, 75A-F6.

The heart of the 1937 story rested in two overlapping themes. One was the continuing ambitions of the Federal Trade Commission to retain sole jurisdiction over regulation of advertising. The second was the political antipathy and lack of communication between the New York Senator and Franklin Roosevelt.

Many people in the affected industries were convinced that the political factor would be crucial insofar as passage of S. 5 was concerned. In November 1936, *Drug and Cosmetic Industry* reported that New Dealers had no intention of passing a new food law which would bear Copeland's name. For that matter, the periodical continued, the "inside dope" was that except for the death of House Speaker Byrns in June 1936, S. 5 would never have reached the floor for a vote in the lower chamber. Byrns had determined to block S. 5 because of Copeland's deviations from the New Deal fold.[48]

Whether this report was correct may be debatable but that FDR's relations with Copeland were affected by their political differences is quite clear. The President's reply to his friend Dr. Harvey Cushing in December 1936 leaves little room for doubt. Cushing wrote Roosevelt to urge his support for Copeland's effort to lodge advertising powers with FDA. "My difficulty," the President replied, "has been a political one. I hope to have someone other than your medical colleague, Dr. Copeland, handle it [the food-drug bill] at the coming session. Enough said!"[49]

The *Washington Daily News* reported in January 1937 that Copeland was ready to "bury the hatchet" with FDR in the light of the popular mandate which the New Deal had received in the fall of 1936. If so, the Senator had several recent political sins for which to atone. He had refused to go to the National Democratic Convention. He had taken the occasion of the celebration marking the 150th

[48] "Reviving the Drug Act," *D&C Industry* 39 (November 1936), 583, 609.
[49] Roosevelt to Cushing, December 28, 1936, PPF 1523, Roosevelt Papers.

year of Tammany Hall to verbally flay the administration. He had gone off on a visit to the Holy Land during the fall election campaign, while at home his wife spoke in support of Alf Landon. The *Daily News* was sure FDR was not inclined to overlook these sins.[50]

Nor had the Senator's relations with his New Deal colleagues or the President been enhanced by his recent personal activities. The press had lately charged that Copeland used his position as Chairman of the Commerce Committee to acquire free ship passage to Bermuda.[51] The old charges that the Senator in his private business dealings was financially tied with the patent medicine industry continued to be made.[52] Whatever Copeland's feelings about the ill will of FDR, the stains upon his personal reputation were apparently a matter of some concern by 1937. In that year the Senator published a small booklet entitled *Toward Better Understanding*,[53] which was an across-the-board defense of his financial and personal business dealings.

If Copeland truly wanted to end his rift with Roosevelt his conduct was a classic study in ineptitude. With the beginning of 1937, he threw himself, full fury, into the campaign against FDR's court revision plan.[54] Indeed, he became one of the President's most vocal opponents. *Business Week* felt that the Senator's court posture might well doom the new drug bill. FDR would block his pet measure in retaliation.[55] At the moment, however, Copeland had in a sense incorporated the coming court battle into his food-drug bill strategy. He intended to rush S. 5 out of committee in the lull before the Senate entered the court controversy. The lull would provide hopefully the psy-

[50] January 4, 1937, FDA Scrapbooks, Vol. 14.
[51] *Ibid.*
[52] Consumers Union *Reports*, March 1937, *ibid.*, Vol. 15.
[53] Located in Bailey Papers.
[54] *Washington Star*, February 24, 1937, FDA Scrapbooks, Vol. 14; radio address by Copeland, March 12, 1937, Copeland Papers. Copeland's successive statements on the court issue may be followed in Copeland Scrapbooks, Vols. 33-35.
[55] "F.D.R. Jars Food Bill," *Business Week*, February 27, 1937, 16.

chological moment to gain quick passage.[56] He was able to avoid new public hearings—much to the irritation of the women's organizations—and get S. 5 reported out on February 15, 1937.

The haste was costly—too costly for many consumer groups. As FDA's Ruth Lamb put it, the "well known wrecking crew of Bailey, Clark and Vandenburg" had restored the old Bailey amendment "in all its pristine glory."[57] Multiple seizure of misbranded items was once more confined to material "imminently dangerous to health," and the following phrase, "or is, in a material respect, false, grossly misleading or fraudulent" had been omitted.[58] The weakness of control over misbranded items was compounded further in that the definition of misbranding was more restricted. Labeling must be false or misleading in some "material" particular. The word "material" was an innovation.[59] FDA was not happy with these changes.[60] The women were apoplectic. Still, with all, S. 5 would doubtless have passed the Senate in this form if not for the unexpected intervention of President Roosevelt.

FDR had heard of the changes before the committee report. On February 16, he wrote a brief note to majority leader Joseph Robinson enclosing an FDA critique of the alterations. The President argued that the idea of a new law was, after all, to strengthen control over that of the 1906 statute and "the Copeland bill produces, I fear, a contrary result."[61] Robinson sought to minimize the importance of the changes in his reply to the President. Insertion of the word "material" was not a significant matter. The

[56] "Quick Thinking," *PI* 178 (February 18, 1937), 118.
[57] Lamb to Louis Engels, February 17, 1937, Correspondence, Box 863.
[58] *Ibid.* The full text of the Senate version may be conveniently found in Charles W. Dunn, *The Federal Food, Drug, and Cosmetic Act*, 636-75. On the above point see 662.
[59] *Ibid.*; see also 668.
[60] Crawford to Alan Perley, March 2, 1937, Commissioners' File, Box 12.
[61] Roosevelt to Robinson, February 16, 1937, OF 375, Roosevelt Papers.

Bailey amendment, he admitted, was harmful, but Robinson felt it would be next to impossible to strike it from the bill. Besides, the majority leader concluded, S. 5 had already been reported out of committee.[62] In this case Roosevelt remained unmollified and undaunted. At a press conference on February 23, and apparently without warning to Copeland, he lashed out at S. 5. He charged that the bill was weaker than the present Wiley law and hinted that if the bill reached his desk in its present form he would use the veto.[63] It was hardly surprising that on February 24 Copeland asked the Senate to postpone a vote on his measure and invited new amendments.[64]

To *Business Week* an obvious factor in the President's verbal blast was his political irritation with Copeland. FDR had gone out of his way, the periodical maintained, to emphasize the weaknesses rather than the strengths of the bill.[65] The charge very likely had some substance in the light of the President's past lack of interest in the measure, at least before the public. Yet this was not the whole story. It was widely acknowledged in the press that the President's action had been prompted by the urgings of his wife. Mrs. Roosevelt in turn was responding to the pleas of the national women's organizations.[66] When Copeland had failed to fight, the women went directly to Eleanor and quite obviously with good results from their point of view.

Printers' Ink felt that the women and "the crusading clique in the Department of Agriculture" would have cause to regret the move. The President's "sarcastic and enigmatic" comments on S. 5 had, that journal contended, put

[62] Robinson to Roosevelt, February 19, 1937, OF 375, Roosevelt Papers.
[63] *New York Herald-Tribune*, February 24, 1937, FDA Scrapbooks, Vol. 14; "The Week," *New Republic* 90 (March 24, 1937), 195; *New York Times*, February 24, 1937, FDA Scrapbooks, Vol. 14; "Food and Drug Free-for-All," *PI* 178 (February 25, 1937), 12, 14.
[64] *Ibid.*
[65] "F.D.R. Jars Food Bill," *Business Week*, February 27, 1937, 16.
[66] *Philadelphia Record*, March 11, 1937, FDA Scrapbooks, Vol. 15; "Recent White House S. 5 Stand," *PI* 178 (March 4, 1937), 12; *The Capital Daily*, February 26, 1937, FDA Scrapbooks, Vol. 14; "F.D.R. Jars Food Bill," 16.

the bill out of the running for the time being. Moreover, *PI* continued, FDA may well have prejudiced their chances to be given advertising controls. The Senate had been favorable to FDA and repeatedly beaten off "raiding parties" from FTC. That body would resent this FDA-prompted intervention in their affairs—perhaps enough to take some second looks at the case for the Federal Trade Commission.[67] Whether Copeland liked the backstage maneuver by his allies or not, he gave no indication of anger. Indeed he stated publicly that "the President was absolutely right." He was positive also that the matter could be worked out in a way satisfactory to all concerned very soon.[68]

The "matter" was worked out at least to the "relative" satisfaction of all concerned. On March 2, Copeland requested that the report on S. 5 be returned to the Committee on Commerce. The committee wished to present a revised report which would reflect certain new amendments. The new report was submitted to the Senate on March 8. The bill was debated on March 8 and 9. The debate touched a variety of points but major alterations in the measure were limited to the seizure and misbranding clauses.[69] This time Copeland had chosen to consult FDA and the agency had concurred in the changes made.[70] In the matter of multiple seizure of misbranded items the phrase "imminently dangerous to health" had been altered to "actually dangerous to health." The additional phrase "or is, in a material respect false and fraudulent" omitted in the February draft had been restored. In the misbranding section the word "material" was removed. The definition allowed misbranding charges when labeling was "false or misleading in any particular."[71]

[67] "Recent White House S. 5 Stand," 12.
[68] *The Capital Daily*, February 26, 1937, FDA Scrapbooks, Vol. 14.
[69] The text of the debate may be found in *Congressional Record*, 75th Cong., 1st Sess. (March 8 and 9, 1937), 1961-62, 2001-21.
[70] Crawford to Copeland, March 1, 1937, Petitions and Memorials File, 75A-J12.
[71] For convenience these changes may be located in Dunn, *Federal Food, Drug, and Cosmetic Act*, 780, 786.

Senate bill 5 passed that chamber on March 9 and was sent to the House. There it faced a very uncertain future. Rumors of political intrigue were rampant. The chief one was that Administration forces in the lower assembly would throw out the whole of Copeland's measure and write their own version. The purpose of this strategy was to force ultimately the passage of the House substitute and thus prevent the inauguration of a new food statute which would carry the New York Senator's name.[72]

This rumor was given increased credence in a very few days when Chairman Clarence Lea of the House Commerce Committee announced that his committee did intend to draft its own measure.[73] The tales of political intrigue appeared even more justified in May when Roosevelt was questioned about the present suspended animation of S. 5. There seemed to be a difference of opinion over the location of advertising controls, reporters told FDR, who feigned his usual lack of knowledge on the bill. His enigmatic reply suggested his information was not so limited. "I don't believe the jurisdiction end of it is the real problem," the President answered. "I believe there is something else in it."[74]

Perhaps, in fact, there was "something else in it," but the FTC theme would be the dominant one in the House. The forces of that agency were well deployed by the first of the year and in the coming battle FTC had several valuable assets. For one thing, the Commission could, and did, play the political motif to good advantage. Very early in the new session they were spreading the word that a vote for S. 5 meant "bestowing kudos on a White House foe." A "wily line," columnist Drew Pearson called it.[75] A second decided strategic advantage was, as Ruth Lamb lamented, that most of the commissioners were ex-members of the

[72] "New and Non-Official," D&C Industry 40 (March 1937), 328.
[73] New York Post, March 24, 1937, FDA Scrapbooks, Vol. 15.
[74] Franklin Roosevelt Press Conferences, May 28, 1937, Vol. 9, 404, Roosevelt Papers.
[75] "Washington Merry-Go-Round," Philadelphia Record, December 27, 1936, FDA Scrapbooks, Vol. 13.

House and could thus carry their lobbying activities right to the floor of the chamber among their old colleagues.[76] Even more advantageous, the commissioners did not labor in their lobbying under the threats of reprisal which faced FDA. The provisions of the 1919 Deficiency Appropriations Act forced caution on the Food and Drug Administration. The commissioners were protected by the judicial precedent of the recent Humphrey Executor case. There the court held that members of FTC could not be removed except for specific cause stipulated in the organizational statute.[77]

A fourth advantage was the recent elevation of Clarence Lea to head the House Committee on Interstate and Foreign Commerce when Sam Rayburn moved up to the Speaker's post. Rayburn had been a friend of FDA but Lea was not, as the personnel of the drug unit understood quite well. Charles Crawford wrote David Cavers that Lea had been "downright hostile" toward the desires of the agency to obtain advertising control.[78] *Business Week* felt that at least part of the Chairman's bias was related to the "trouble" which FDA had stirred up over the condition of dried fruit being shipped from Lea's California constituency in recent years.[79] Whatever the particular reason, events would prove that the California Congressman was far less than receptive to the cause of FDA.

In January 1937, Senator Burton Wheeler had again introduced a bill to expand the regulatory powers of the Federal Trade Commission. It bore the designation S. 1077. This bill passed the Senate on March 29, 1937. The Wheeler bill was a general expansion measure and did not seek specifically to vest all drug, cosmetic, and food adver-

[76] Lamb to Meta Pennock, March 20, 1937, Correspondence, Box 863.

[77] "Official Family Feud," *PI* 178 (January 14, 1937), 20-21. For a brief summary of the Humphrey case (1935), see Saye, Pound, and Allums, *Principles of American Government* (Englewood Cliffs, N.J., 1958), 243-44.

[78] Crawford to Cavers, May 22, 1937, Commissioners' File, Box 12.

[79] "Ready for Drug Bill 'Solution,'" *Business Week*, June 12, 1937, 25.

tising authority in the Commission. On March 23, Clarence Lea introduced a House version of the Wheeler measure, however, which was designed, according to *Printers' Ink,* as "the last word on all advertising." The measure placed special emphasis on food, drug, and cosmetic copy control. "It does seem," *PI* editorialized, "as if Santa Claus has really come to the Federal Trade Commission."[80]

Printers' Ink was not happy about the situation. On the occasion of the introduction of S. 1077 that journal stated with reference to S. 5 and Chapman's H.R. 300 that a third bill "is just about as much needed as were Maine and Vermont in the Democratic electoral column." They charged that the Wheeler bill and the coming Lea opus had been promoted purely and simply by the "ridiculous" power cravings of FTC.[81] On into the spring the periodical attempted to rally support for S. 5 and admonished the trade not to be scared off "by the beating of tom-toms by the Federal Trade Commission lobby."[82]

Clarence Lea remained unmoved by such editorial copy. His strategy was calculated and efficient. The first move was to minimize the power of Virgil Chapman. Instead of appointing Chapman to head the subcommittee which would consider S. 5, Lea retained the position himself. The Kentucky Congressman was simply a member of the subcommittee.[83] Lea hedged his bets further, according to Chapman, by "stacking" the subcommittee "100% for FTC."[84] The remainder of the plan was immediately clear enough to many people. The California Congressman would bottle up S. 5 in committee until his own FTC bill was

[80] "New Food and Drug Bill," *PI* 178 (March 25, 1937), 12, 14. For detail provisions of the respective Wheeler and Lea bills, see Charles W. Dunn, *The Wheeler-Lea Act: A Statement of Its Legislative Record.*

[81] "Noisy and Troublesome," *PI* 178 (January 21, 1937), 124-25.

[82] "Why the Lea Bill, Anyhow?" *ibid.,* April 1, 1937, 120-21.

[83] Crawford to Cavers, May 22, 1937, Commissioners' File, Box 12.

[84] Robert Allen to Lamb, July 31, 1937, Correspondence, Box 861; *Advertising and Selling,* July 27, 1937, FDA Scrapbooks, Vol. 16.

reported out and passed the House. Then the Copeland measure would be released, written anew by the committee and *sans* advertising provisions.[85]

Official news of the fate of S. 5 in the House was minimal during the spring and early summer. Lea announced that the committee had, in fact, struck the whole of the Senate measure.[86] The trade press continued to report the aforementioned Lea strategy to give sole advertising authority to the FTC. *Food Industries* added that a second part of the plan would be finally to send the Senate a food measure which, in rewritten form, would carry Chapman's, not Copeland's, name. New Dealers were determined that Copeland should not get lasting credit for a new law.[87] Aside from such rumors, the press could only report that S. 5 for the time "slumbered peacefully" in a committee pigeonhole.[88]

Representatives of professional pharmacy seldom commented on specific provisions in the various Copeland bills. At this juncture, however, they did comment on the question of advertising controls and in such way to suggest division within their ranks. The American Pharmaceutical Association concluded that powers over advertising would probably be placed with the Federal Trade Commission. The Association now took the position that location itself was not a crucial question. The important matter was that the function be carried out effectively.[89] The American Association of Colleges of Pharmacy on the other hand had committed itself in January to the investiture of advertising

[85] "Save the Food and Drug Bill," *Nation* 144 (April 24, 1937), 461-62; Salthe to Copeland, March 12, 1937, Petitions and Memorials File, 75A-F6.

[86] "Food and Drug Act," *Am. Druggist* 95 (March 1937), 66-67.

[87] *Ibid.*; "Who Shall Control Advertising," *Business Week*, April 3, 1937, 42, 44; "Food Bill Prospects," *Food Industries* 9 (May 1937), 276; "Lea Bill to Pass House Soon," *PI* 179 (April 1, 1937), 14.

[88] "Food Bill Prospects," 276.

[89] "Presidential Address by President George Denton Beal," *JAPhA* 26 (November 1937), 1015.

powers with FDA and gave no indication of a change in position.[90]

In general, proponents of a new food law, and particularly those who favored Food and Drug Administration jurisdiction over advertising, remained silent about the current state of affairs. *Printers' Ink* ascribed this phenomenon to word passed from the House that the drug unit must take whatever the lower chamber chose to provide or get nothing. Any intervention in the operations of that body such as occurred in the Senate would promise an unpleasant result.[91] FDA's Charles Crawford concluded by May that even Chapman would not fight against the sentiment of the House. Crawford wrote David Cavers that the Kentucky legislator would not yield on his belief that advertising regulation should not be split between two agencies. If the feeling of the House was largely for FTC, Chapman intended to withdraw his opposition, provided enforcement could be given stronger teeth than the mere use of cease and desist orders.[92]

In mid-May, the House committee began closed door meetings on both the Wheeler-Lea bills and the Copeland-Chapman measures.[93] In late May, a test vote in the committee proved that the defenders of FDA advertising authority were in the minority. There was a substantial split, however, among those who believed drastic criminal penalties should be added to the FTC bill and those who held that the cease and desist order procedure was satisfactory. Chapman assumed the leadership of the former and in his efforts got the backing of the national women's organizations. Lea, it seems, was satisfied with the present enforcement practices of the Commission.[94]

The Kentucky Congressman and his California colleague differed on a second matter more pertinent to the drug

[90] Editorial, *AJPhE* 1 (January 1937), 98.
[91] "Lea Bill to Pass House Soon," *PI* 179 (April 1, 1937), 14.
[92] Crawford to Cavers, May 22, 1937, Commissioners' File, Box 12.
[93] "Food Bill Compromise Near," *PI* 179 (May 6, 1937), 12.
[94] "Lea Bill Progresses," *ibid.*, July 1, 1937, 14; "F.T.C. to Rule Advertising," *ibid.*, July 15, 1937, 29.

bill. Chapman believed that Lea intended to push an emasculated, extremely weak version of S. 5 through committee and then report it out too late in the session for a vote in the House. The idea was to get the drug bill off the committee docket and thus prevent any attempt to draft a stronger bill during the next session. This action would be taken, of course, only after the passage of the Wheeler-Lea measure. Chapman determined that if he could not halt the FTC strategy, then at least the drug bill should be held over till the first of the year when the committee would perhaps go along with his efforts to strengthen the whole measure.[95]

The Kentucky Congressman was sufficiently certain of his thesis to carry the case to Roosevelt. If FDR would support him and put in a word with Speaker Sam Rayburn, S. 5 might be held in the committee. While the substance of the conversation is unknown, the President did discuss the food bill by phone with Rayburn.[96] If Chapman was right in his diagnosis of Lea's plan, then the Kentuckian won a limited victory. S. 1077 was reported to the House on August 19, 1937. It provided the FTC with the long-debated control over advertising of food, drugs, and cosmetics. About the same time, however, Lea announced that the bill proper would be held over in committee for further consideration.[97]

In March 1937, the *New York Times* printed a comment from a 1906 copy of *Life* on the status of the Wiley bill. "Who is that shabby looking, patched-up individual trying to get on the floor of the House?" the journal asked. "That's old Pure Food Bill. When he first came here he looked pretty good, but now he has been knocked around and changed so much that his former friends don't know him at all. In a minute you'll see him thrown out bodily again."[98] How apropos the reprint seemed for the end of August.

[95] M.H.M. (Marvin McIntyre) to Roosevelt, August 5, 1937, OF 375, Roosevelt Papers.
[96] *Ibid.*
[97] *Drug Trade News*, August 30, 1937, FDA Scrapbooks, Vol. 16.
[98] *New York Times*, March 14, 1937, *ibid.*, Vol. 15.

Perhaps the current bill had not been "thrown out" but the future seemed dim. Without a drastic change in the course of events proponents of revision might well wish it had been "thrown out." "Dr." Virgil Chapman's wishful prognosis to the contrary, S. 5 was dying.

VII

DOCTOR MASSENGILL'S ELIXIR

Let us hope that at last the public will demand of
Congress that such a law be passed that these more
than ninety innocent victims . . . shall not have died
in vain.

> GENERAL BULLETIN OF CONSUMERS'
> RESEARCH
> JANUARY 1938

The only way in which the industry can properly
protect itself against another accident of the same
type is for an outside agency, such as the government, to have proper control over manufacturers
and their products.

> DRUG AND COSMETIC INDUSTRY
> NOVEMBER 1937

THE summer of 1937 must have been a depressing one for
Royal Copeland. Four years had elapsed since the Senator
introduced S. 1944, and it would appear that final passage
of a drug bill was not even close. All Copeland could do
was to sit and wait to see how the House would handle
S. 5 number 2. The Senator could not know that the fall of
1937 would be decisive in the question of passing a new
law. *Drug and Cosmetic Industry* could not foresee this
either, and, in retrospect, there was much that was ironic
about its May comments on the progress of S. 5.

D&CI felt self-righteous. The industry had been demonstrated correct in its fight on at least one provision of the
old Tugwell measure. That bill listed thirty-six diseases for
which therapeutic claims in medical advertising were forbidden. In the present bill the number of diseases had been
reduced to six. "The manner in which the industry has at
least been proven partially right," asserted *Drug and Cos-*

metic Industry, "is through the discovery of Prontosil, Prontylin or Sulphanilamide [*sic*]. These products will cure at least five of those diseases for which the original Tugwell Bill claimed there was no cure." "Just some more proof," the periodical concluded, "that the Food and Drugs Law must be flexible and allow some leeway to manufacturers."[1]

The trade journal was certainly right that the decade had brought important therapeutic discoveries. Sulfanilamide was an apt case in point. This was the name of one of a group of closely related chemicals first reported in European medical literature about 1935.[2] Sulfanilamide was showing dramatic curative effects in the treatment of a number of infections including gonorrhea, septicemia, and sore throat. American physicians were quick to recognize the far-reaching effects of the drug, and by 1937 its use had grown to tremendous proportions.[3] Pharmaceutical manufacturers rushed to fill the demand—rushed too quickly in the case of at least one concern.

The Samuel E. Massengill Company of Bristol, Tennessee, had been in operation since 1897. The concern was not one of the giants of the medicine industry, but its reputation was solid. Massengill had run afoul of the Food and Drugs Act once or twice, but this was not an unusual situation in his trade.[4] The company's chief chemist was Harold C. Watkins. He received the Ph.C. degree from the University of Michigan in 1901, and joined the Massengill Company in 1935. Prior to that date he had worked for a variety of pharmaceutical concerns and for a while was self-employed. He too had had a minor brush with the law. In 1929 the Watkins Laboratories was cited by the Solicitor of the Post Office Department in regard to a weight reducer pro-

---

[1] "New and Non-Official," *D&C Industry* 40 (May 1937), 632.
[2] *Report of the Secretary of Agriculture on Deaths Due to Elixir Sulfanilamide–Massengill*, U.S. Congress, Senate Documents 124, 75th Cong., 2nd Sess. (1937), 1. Hereafter cited as *Secretary's Report*, Senate Documents 124, 75th Cong., 2nd Sess.
[3] *Ibid.*
[4] *Ibid.*, 2, 8.

duced by the company. The case was dropped, however, when Watkins agreed to abandon the sale of the product.[5]

Among other drugs, the Massengill Company manufactured Sulfanilamide, which was delivered to the public in capsule and tablet form. By June 1937, however, detail men were reporting a substantial demand for a product in a liquid form. Thus far no one in the industry had been able to find a suitable fluid vehicle in which sulfanilamide would dissolve.[6] Watkins took a vacation in June and upon his return to work set about the task of finding a solvent. He tried a number of chemicals and at last hit upon diethylene glycol. Watkins found that he could dissolve as much as seventy-five grains of sulfanilamide in an ounce of the solvent. He decided ultimately on a mixture of forty grains per fluid ounce since in the higher amount ingredients tended to separate out on chilling.[7]

By Watkins' later admission, no tests were made to determine the toxicity of either the separate ingredients or of the finished product. The concoction did pass through a so-called control laboratory but there the "elixir" was merely checked for appearance, flavor, and fragrance. Dr. Massengill, when questioned by FDA inspectors, confirmed Watkins' statement that no clinical tests for toxicity were made. On August 28, 1937, the formula for "Elixir Sulfanilamide" was sent to a second Massengill plant at Kansas City where forty gallons were subsequently manufactured. An initial amount of forty gallons was produced at Bristol followed by two additional batches of eighty gallons each. All told, 240 gallons were manufactured. Distribution to the public began on September 4, 1937.[8]

With Senate bill 5 locked up in the House of Representatives as summer gave way to fall, the scene was set for one

[5] *Ibid.*, 8-9; FDA File on Sulfanilamide–S. E. Massengill Company, AF 1-258, Vol. 1, Federal Records Center, Alexandria, Va. Hereafter cited FDA File on Sulfanilamide-Massengill.

[6] "The Deadly Elixir," *D&C Industry* 41 (November 1937), 611.

[7] FDA File on Sulfanilamide-Massengill, Vol. 1.

[8] *Ibid.*; *Secretary's Report*, Senate Documents 124, 75th Cong., 2nd Sess., 3.

of the nation's worst drug disasters. The story began to unfold on October 11. The Chicago offices of the American Medical Association received a telegram from Dr. James Stephenson, President of the Tulsa County Medical Society, Tulsa, Oklahoma. Stephenson reported that six deaths had occurred in his area among patients taking Elixir Sulfanilamide–Massengill. He wanted to know the composition of the product, which in line with the current drug statute was not given on label material. In response to the telegram the AMA could only inform the Tulsa physician that no product of the S. E. Massengill Company stood accepted by the Council on Pharmacy and Chemistry and that the Council had not recognized any solution of sulfanilamide.[9]

The AMA Chicago office immediately requested specimens of the elixir from Tulsa. A telegram was sent also to Massengill's Bristol plant asking for the composition of the product. Massengill complied but requested that AMA officials keep the information confidential. Subsequent tests by the Chicago medical laboratories indicated that diethylene glycol was a toxic agent. On October 18, the AMA released a general warning to the public via the press and radio.[10]

In the meantime the Food and Drug Administration had become aware of the matter. First word of the Tulsa deaths reached FDA on October 14 through a telephone call from a New York physician associated with a large drug manufacturer. Presumably he had acquired his information from trade contacts in the Tulsa area. A full investigation was ordered through the Kansas City station. The following day an FDA inspector arrived in Tulsa. He found that Dr. Stephenson had already asked the city's Retail Druggist Association to hold up the sale of Elixir Sulfanilamide pending further investigation. The inspector was informed further that Tulsa suspicions about the drug were based upon

9 "Elixir of Sulfanilamide-Massengill," *JAMA* 109 (November 6, 1937), 1531.
10 *Ibid.*

the clinical histories of nine children treated for strepto-
coccic infections and one adult treated for a gonorrheal
infection.[11]

By the time of the inspector's preliminary report to
Washington on October 17, seven children and two adults
were dead. The toll might be expected to rise since some
thirty prescriptions for the drug had been authorized before sales were halted. The FDA agent also attended an
autopsy on the most recent adult fatality. The patient had
become anuric in a Tulsa hospital to which he had been
admitted. Kidney function was finally restored, but the
following day complete anuria occurred once more and
death followed. The autopsy showed that the patient's
kidneys were enlarged 50 percent over normal, and were
clotted and purplish in color.[12] As deaths mounted across
the nation in the next weeks, a pattern of symptoms for
Elixir Sulfanilamide victims began to be clear. The painful
nature of their deaths was equally apparent. Victims were
generally ill from seven to twenty-one days. They suffered
intense pain. Common symptoms were stoppage of urine
and severe abdominal pain. Nausea, vomiting, and convul-
sions preceded death in some cases.[13]

According to the report of one physician, the onset of
symptoms in his patients ranged from immediate reaction
to the first dose of the elixir to a maximum of six hours.
Abdominal pain came so quickly and with such intensity
that several physicians recommended immediate surgical
intervention to their patients.[14] Curiously, however, the
effect of the drug varied a good deal among takers. Many
persons who took it but discontinued use with the onset
of the symptoms completely recovered. One individual took

[11] FDA Chronological File of Reports on Deaths from Elixir Sulfanilamide, AF 1-258, Sub-file 510-.2055, FRC, Alexandria, Va. Hereafter cited FDA Chronological File on Deaths.

[12] *Ibid.*

[13] *Secretary's Report,* Senate Documents 124, 75th Cong., 2nd
Sess., 7; "Deaths Due to Elixir of Sulfanilamide Massengill," *JAMA*
109 (December 11, 1937), 1987.

[14] FDA Chronological File on Deaths.

over seven fluid ounces with no ill effect. One child died from less than two fluid ounces.[15]

While Food and Drug officials pursued the matter of the Tulsa deaths, other FDA inspectors rushed to the Kansas City and Bristol plants of the Massengill Company. On the morning of October 18, federal agents interviewed Massengill and Watkins. It was on this occasion that the two men admitted the failure to perform toxicity tests. Watkins stated, however, that since the news of the fatalities he had personally taken four ounces of diethylene glycol and teaspoonful doses of Elixir Sulfanilamide with no adverse effect. He stated further that tests of the solution had now been run on guinea pigs with no fatal results. Massengill told the FDA agents that he believed the Tulsa deaths might have been the result of using the elixir in combination with other drugs.[16]

Pathetically, on October 20 Massengill wired the American Medical Association asking if that organization could prescribe an antidote for his own solution. The AMA replied that none was known.[17] The same day the press reported Walter Campbell's latest statement on the growing tragedy. By that time fourteen people were dead. Campbell admitted that as yet there was no clear answer to the cause of the diaster though it was believed that the diethylene glycol was responsible. "We do know," the drug chief emphasized, "that there was something radically wrong."[18]

The "wrong" was more than a matter of Massengill's Elixir. Because of the limits of the 1906 law, as Campbell made clear, FDA was forced to proceed against the lethal medicine on technical grounds of misbranding.[19] Since the law had no provision against dangerous drugs as such,

[15] *Secretary's Report*, Senate Documents 124, 75th Cong., 2nd Sess., 7.
[16] FDA File on Sulfanilamide-Massengill, Vol. 1.
[17] "Elixir of Sulfanilamide-Massengill," *JAMA* 109 (November 6, 1937), 1539.
[18] *Baltimore Sun*, October 20, 1937, clipping in FDA File on Sulfanilamide-Massengill, Vol. 1.
[19] *Ibid.*

seizure had to be based on a charge that the word "elixir" implied an alcoholic content. Massengill's product was a diethylene glycol solution. The law was thus violated since the Pharmacopeia defined elixirs as containing alcohol, and this medicine did not. If the product had been called a "solution" rather than an "elixir" no charge could have been brought.[20]

Whatever the legal technicality, FDA entered the case with a fantastic zeal and efficiency. Practically the entire field force of 239 Food and Drug Administration inspectors and chemists were assigned to the work. They in turn were heavily supported by state and local health authorities. The magnitude of the job ahead only gradually became evident. For one thing, valuable time was lost, and the danger of further fatalities increased because of the nature of Massengill's initial efforts to recall unsold quantities of his product from the market. Telegrams sent out to jobbers, pharmacists, doctors, and branch houses on October 15 asking immediate return of stocks gave no indication of the dangerous nature of the product. On October 19, FDA inspectors assigned to the various offices of the company insisted on a new round of telegrams specifically stating that the product might be dangerous to life.[21]

Yet such momentary difficulties with Dr. Massengill were minor matters compared to the complexity of the seizure problem itself. Indeed, it should be emphasized that the doctor was, on the whole, quite cooperative with government agents. The real job before drug officials was to follow through on all shipment records and consignee's records as well as contacting two hundred Massengill salesmen to locate physician's sample bottles. The work was complicated by the fact that while the elixir was a prescription product some doses had passed over drugstore counters without prescription, largely because the public had

[20] *Secretary's Report*, Senate Documents 124, 75th Cong., 2nd Sess., 9.
[21] *Ibid.*, 4-5; "Deaths Due to Elixir of Sulfanilamide-Massengill," *JAMA* 109 (December 11, 1937), 1986.

learned that sulfanilamide could cure gonorrhea. In a few cases the druggists had no full record on the recipients.[22]

While all concerned in the seizure operation were generally quite responsive to drug officials, some uncooperative reactions, perhaps from fear, inevitably did occur. One Massengill salesman in Texas was so thoroughly hostile to demands for information that he was put in jail by state authorities before he decided to reveal the needed data.[23] Fearing for personal reputation, one doctor removed and destroyed a prescription from a drugstore file. A druggist scratched the names of patients off the prescription forms in his records. In South Carolina a physician flatly denied prescribing the elixir to a Negro patient who had recently died. The inspector pursued the matter with the victim's family. He was told that it was the custom in the area to place the personal effects of the deceased on the grave. The FDA agent set out over dirt roads to the backwoods burial ground. On top of the grave lay a four-ounce bottle of Elixir Sulfanilamide with three ounces missing.[24]

Most physicians and pharmacists were too shocked by the disaster to even consider withholding aid. Typical was the attitude expressed in writing by Dr. A. S. Calhoun of New Orleans:

> . . . to realize that six human beings, all of them my patients, one of them my best friend, are dead because they took medicine that I prescribed for them innocently, and to realize that medicine which I have used for years in such cases suddenly had become a deadly poison in its newest and most modern form, as recommended by a great and reputable pharmaceutical firm in Tennessee; well that realization had given me such days and nights of mental and spiritual agony as I did not believe a human being could undergo and survive. I have spent

[22] *Secretary's Report,* Senate Documents 124, 75th Cong., 2nd Sess., 6-7.
[23] *Ibid.*
[24] FDA Chronological File on Deaths.

hours on my knees. . . . I have known hours when death for me would be a welcome relief from this agony.[25]

And well might physicians, as well as common citizens, be shocked. The reports of fatalities continued to pile up. A small child died in Arkansas, the first victim in that state. After several days of agony a twenty-five year old Negro man died in a Vicksburg hospital. He had taken four ounces of the elixir. In San Francisco physicians fought a losing battle to restore the kidney functions of a six-year-old treated with the drug for an infected throat. During September and October 1937, seventy-three people died as a direct result of using the solution. It was suspected that the deaths of an additional twenty people were in whole or part brought about by the drug. Fatalities took place in fifteen states, as far east as Virginia and as far west as California.[26]

The FDA seizure drive was made more complicated by the tide of public fear as news of the disaster spread. Letters to FDA, such as one from a frightened man in Coral Gables, Florida, were common. Correctly or not, he believed that he had taken doses of the elixir. An inspector was assigned to investigate.[27] Every rumor had to be cleared. There were also the moments of deep pathos. At one station a small boy of nine or ten, dressed in faded clothes, came in bearing a bottle labeled Elixir Iron, Quinine, and Strychnine Phosphate. The boy, like the man from Florida, was frightened. He had heard the stories of the deaths from Elixir Sulfanilamide and had fearfully noted that his mother's medicine bore the word "Elixir." His

[25] *Secretary's Report*, Senate Documents 124, 75th Cong., 2nd Sess., 6-7.
[26] Article in *Fort Smith* (Ark.) *Southwest American*, October 28, 1937; R. D. Sherman to Chief, Central District, October 29, 1937; Bernard M. Sorauf to M. L. Yakowitz, October 22, 1937, all in FDA Chronological File on Deaths; *Secretary's Report*, Senate Documents 124, 75th Cong., 2nd Sess., 1.
[27] E. G. Pierce to FDA, Washington, November 5, 1937; G. P. Larrick to Chief, Atlanta Station, November 10, 1937, both in FDA File on Sulfanilamide-Massengill, Vol. 1.

relief was profound when informed that the medicine he brought in was not the fatal compound.[28]

As the death toll mounted and the heated search for outstanding amounts of the sulfanilamide solution became more frantic, the FDA neglected no leads, no matter how far-fetched they might appear. New York complaints about a colloidal sulphur drug produced by Massengill precipitated a rush investigation in that state.[29] Rumors that another Massengill product called Aspiral included diethylene glycol sent Central District inspectors dashing to investigate and to begin seizure operations if the rumor proved to have substance.[30] In Texas there was a sudden flurry of activity over Elixir Creozote produced by the Bristol firm. It had come to the attention of Texas health officials that some patients experienced nausea reactions from the drug. FDA was notified and an investigation began at once.[31]

The most amazing facet of the entire sulfanilamide excitement, however, was the efficiency of the Food and Drug Administration. Under its leadership 99.2 percent of the original 240 gallons of the fatal drug was accounted for. The total of 107 deaths had been the result of a public consumption of about six gallons. Had the total 240 gallons been consumed, and fatalities occurred in the same ratio, over 4,000 Americans would have died.[32] The weakness of the operation was the pitiful legal powers which FDA possessed in its attempts to safeguard the public. It is noteworthy that part of the agency's interest in Massengill's Elixir Creozote was to strengthen its sulfanilamide case. The agency was seeking some additional violative counts to use because of the "grave possibility," as one FDA official put it, that a legal violation could not be proved on

[28] *F&D Review*, December 1937, FDA File on Sulfanilamide-Massengill, Vol. 1.
[29] *Ibid.*
[30] *Ibid.*
[31] *Ibid.*
[32] Pamphlet, Fred B. Linton, *Federal Food and Drug Laws— Leaders Who Achieved Their Enactment and Enforcement* (Chicago, n.d.), 68.

sulfanilamide alone.[33] That the death of 107 Americans was not *sure* grounds for prosecution was a telling commentary of the adequacy of the 1906 statute.

The limited nature of the current law was underscored by Massengill himself. In a November letter to the American Medical Association the drug manufacturer expressed regret over the disaster but added, "I have violated no law."[34] In fact, the most that the government did do ultimately was to secure Massengill's conviction on a long list of misbranding counts. The manufacturer paid a $150 fine for each count.[35] The total amounted to $26,000 and it represented the largest fine ever levied under the 1906 law.[36] In the meantime, the doctor sought to defend his actions as best he could. By late November he was circularizing his colleagues and customers with a question and answer brochure entitled *The Facts about Elixir Sulfanilamide.*[37]

In this document the Bristol manufacturer denied any error in compounding the Elixir Sulfanilamide formula. Massengill now insisted that tests had been run on the elixir before it went on the market. No bad results had been indicated. He suggested that there remained much doubt about whether diethylene glycol was, after all, the toxic agent. Perhaps most significant of all in the long run was one question which had little to do with the sulfanilamide disaster. Had his company opposed enactment of a new food and drug statute? Massengill answered emphatically that the concern had never opposed a new law.[38]

The Bristol drug manufacturer well understood that a solid connection had been forged already between Elixir Sulfanilamide and the fate of Senate bill 5. If he had any

[33] FDA File on Sulfanilamide-Massengill, Vol. 1.

[34] "Deaths Due to Elixir of Sulfanilamide-Massengill," *JAMA* 109 (December 11, 1937), 1987.

[35] Article, *Kansas City Journal-Post*, October 19, 1938, FDA File 40564, Seizure Number 21573-C, FRC.

[36] *Ibid.*; James Harvey Young, "The Elixir Sulfanilamide Disaster," *Emory University Quarterly* 14 (December 1958) 234.

[37] Photostat pamphlet in FDA File on Sulfanilamide-Massengill, Vol. 1.

[38] *Ibid.*; "Post-Mortem," *Time* 30 (December 30, 1937), 48-49.

doubts he need only read the December 11, 1937, issue of *JAMA*. That journal's report on the disaster concluded: "It is worthy of note that, shocking as these instances have been, the actual toll in deaths and permanent injury from potent drugs is probably far less than that resulting from harmless nostrums offered for serious disease conditions." Four legislative recommendations followed.[39]

Press coverage of the drug disaster almost of necessity became part of the larger dialogue regarding a new food and drug bill. The *New Orleans Picayune* discussed the matter in terms of what would have been the case if a new law had been on the books.[40] The *Sacramento Bee,* under the caption "Laws Controlling Sale of Drugs Need Tightening" told the story of a grief stricken mother whose young daughter had died from the elixir.[41] In its November 1 issue, *Time* commented on the sulfanilamide situation and recalled that in 1930 many people died from taking a tonic called Jamaica Ginger. The drug bill before Congress would make producers responsible for such blunders, *Time* pointed out.[42] The *New Republic* noted: "If there were an adequate food-and-drug law in the United States it is unlikely that an accident of this sort could have happened."[43]

Mrs. Maise Nidiffer of Tulsa wrote a letter to President Roosevelt. In pathetic tones she spoke of her six-year-old daughter Joan, who had recently died from using the elixir. She recalled that for nine days she had watched Joan's "little body tossing to and fro" and heard "that little voice screaming with pain." "Tonight, President Roosevelt, as you enjoy your little grandchildren of whom we read about," Mrs. Nidiffer continued, "it is my plea that you will take steps to prevent such sales of drugs that will take little

[39] "Deaths Due to Elixir of Sulfanilamide-Massengill," *JAMA* 109 (December 11, 1937), 1988.
[40] *New Orleans Picayune*, November 17, 1937, Correspondence, Box 862.
[41] *Sacramento Bee*, December 2, 1937, Correspondence, Box 862.
[42] "Fatal Remedy," *Time* 30 (November 1, 1937), 61.
[43] "Drugs on the Market," *New Republic* 92 (November 3, 1937), 354.

lives and leave such suffering behind and such a bleak out-
look on the future as I have tonight."[44] The widely distributed story could not help but be mighty ammunition in the
fight for a new drug law.

The food-drug bill dialogue had gone on for four long
years. The sulfanilamide disaster, however, brought a new
emotional content to the matter. The dialogue acquired
new urgency; old positions were shaken and, above all,
more voices were added than ever before. For the first
time in four years the national press brought the matter in
a meaningful way to the public, and the public responded.
Staunch opponents of strong legislation such as Josiah Bailey found his mail heavy with demands for a new law "so
to make impossible a repetition of the recent Sulfanilamide tragedy."[45] Members of the House got the same message. The past legislative apathy toward the bill which
characterized some Congressmen was illustrated in the
person of Representative Frank Crowther of New York.
Crowther had received a letter from a constituent demanding to know why a new bill had not passed. He sent the
letter to Copeland asking if the Senator had an answer.
Copeland wrote back that S. 5 had "passed the Senate a
year ago and we are awaiting House action. Why can't
you hurry it out?"[46]

As the sulfanilamide death toll rose old voices in the
food-drug bill fight became louder and more positive. Consumers' Research flatly placed the blame for the fatalities
on the shoulders of "politicians and others who defeated
some of the best of the earlier food and drug law proposals."[47] The medical profession took a firmer stand than
ever before. *JAMA* stated that the Food and Drug Administration in its efforts to protect consumers was "as ineffi-

[44] *Secretary's Report*, Senate Documents 124, 75th Cong., 2nd
Sess., 8. See also FDA Scrapbooks, Vol. 16.

[45] For example, see Mecklenburg Medical Association to Bailey,
March 20, 1938, Bailey Papers.

[46] Copeland to Frank Crowther, January 1, 1938, U.S. Senate, Committee on Commerce, Petitions and Memorials File, 75A-F6.

[47] "Elixir Poisonings," *General Bulletin* 4 (January 1938), 9.

ciently armed as a hunter pursuing a tiger with a fly swat-
ter." That journal repeatedly criticized Massengill's lack of
precautions and called for a law which would force "common scientific decency" on drug manufacturers.[48] *Hygeia's*
December issue summarized the story of the drug disaster
and called for the passage of drug legislation pending in
the Congress.[49] The December *American Journal of Public
Health* echoed the demand for passage of the Copeland
measure.[50] Old allies of legislative reform such as the national women's organizations and the more professional
pharmaceutical groups of course reacted to the tragedy
with new calls for immediate Congressional action.[51]

The drug trade felt the pressure of the elixir scandal.
Some journals, like the *American Druggist,* fell strangely
silent, offering no coverage or comment on the whole episode.[52] Some proprietary manufacturers, like Lee Bristol,
sought to keep their own skirts clean by pointing out that
the elixir was after all an ethical product. "Manufacturers
of proprietary drug products," Bristol told the National
Association of Insecticide and Disinfectant Manufacturers,
"favored drastic control over untried potentially dangerous
drugs." No group, he continued, exercised more care in
their products than the well-known nationally advertised
remedies.[53]

Drug and Cosmetic Industry spoke for the majority of
the trade, however, when it insisted that a new law was
now an absolute necessity. The elixir tragedy was a good

[48] "Deaths Following Elixir of Sulfanilamide-Massengill," *JAMA*
109 (November 6, 1937), 1545.
[49] "Elixir of Sulfanilamide Deaths and New Legislation," *Hygeia* 15
(December 1937), 1067.
[50] "We Need a New and Strong Food and Drugs Act," *Am. Jnl. of
Public Health* 27 (December 1937), 1286.
[51] See for example, "The Elixir Sulfanilamide Case," *Jnl. of Home
Economics* 30 (February 1938), 109; "The Quick and the Dead,"
AJP 109 (October 1937), 492-94; "The Elixir of Sulfanilamide Experience," *JAPhA* 26 (December 1937), 1225-26.
[52] A survey of the issues of *Am. Druggist* from the fall of 1937 to
mid-1938 failed to locate any coverage or comment on the sulfanilamide affair.
[53] "Proprietaries and the Elixir," *D&C Industry* 41 (December
1937), 778.

example of what could occur when drug manufacturers tried to "outsmart" one another in a scramble to produce an improved product. The incident in itself had severely hurt public and medical confidence in the industry. The industry as a whole must protect itself from a recurrence. That meant the passage of a new drug law. The journal called upon the federal administration to lay aside political infighting so as to get a new law with maximum speed.[54]

At the same time *Drug and Cosmetic Industry*, in writing of its fear about the future, probably also spoke for the majority of the trade. The periodical lamented that there was a great danger in the current wave of sympathy for those who died. Under such pressure the Congress might pass a "far too stringent" bill. Those who had previously stood up for the industry in Congress would now be quite shy about doing so. "The type of law which might be enacted," the periodical stated, "sort of takes the joy out of getting this problem solved."[55]

Not the least of those who clearly understood the useful force of the elixir tragedy were Food and Drug Administration officials. *Newsweek* declared that FDA was "jubilant" over the fact that the sulfanilamide product had been produced in the Tennessee constituency of Representative Carroll Reece, one of S. 5's strongest opponents.[56] Perhaps the word "jubilant" was too strong, but agency spokesmen like Ruth Lamb had no reservations about mentioning Reece's connection in her official correspondence.[57] Nor did the drug unit fail to make extensive use of printed matter, periodical and official, on the disaster to woo new support for a law.[58]

The sulfanilamide episode even attracted the attention

[54] "The Deadly Elixir," *ibid.*, November 1937, 611; "The Elixir and the Industry," *ibid.*, 614.
[55] "New and Non-Official," *ibid.*, 612.
[56] *Newsweek*, November 22, 1937, FDA Scrapbooks, Vol. 17.
[57] Lamb to Helen Woodward, January 31, 1938, Correspondence on Legislation, Carton 125, Accession 52-A89, RG 88, FRC, Alexandria, Va. Hereafter cited as Correspondence, Carton——, Acc. 52-A89, RG 88, FRC.
[58] See Lamb to Dr. Henry Christian, December 14, 1937, Correspondence, Box 859.

of Hollywood's movie moguls. With the permission and, one might assume, the encouragement of FDA, two short movies, entitled *Permit to Kill* and *G-Men of Science* went into production. At the end of each, Walter Campbell appeared and talked briefly on the need for an adequate drug law.[59] Ruth Lamb wrote in delight to a friend that 20th-Century-Fox was planning to make a movie version of the *American Chamber of Horrors.*[60] "Things look better and better for some action," she wrote another friend. "This sulfanilamide situation, while sad enough for the victims, looks as though it would result in their not having suffered for nothing." Even *Drug and Cosmetic Industry* was panning government red tape, she said, "certainly a change of front to be noted."[61]

There were those who were very unhappy about FDA's turning the elixir tragedy into a weapon for reform. Old agency enemies like Howard Ambruster of earlier ergot fame were incensed. The agency's claim that it could not deal adequately with the Massengill Company under the present law, Ambruster charged in a letter to Henry Wallace, was simply a coverup to turn aside public indignation at lax practices in FDA.[62] Proponents of a new law were pleased with the drug unit's handling of the elixir case. The Seattle Commissioner of Health, Frank Carroll, as a case in point, wrote to Copeland urging him to "make all the use you can of the recent sulfanilamide tragedy."[63]

The Congress was in special session during the fall of 1937. That body did not, of course, need Royal Copeland to tell them that the sulfanilamide episode was serious business or that it had some bearing on the fate of S. 5. On November 24, Representative Rees of Kansas made this fact

[59] *Drug Trade News*, November 8, 1937, FDA Scrapbooks, Vol. 17.
[60] Lamb to William Tiedt, December 23, 1937, Correspondence, Box 863.
[61] Lamb to Alice Wright, November 30, 1937, Correspondence, Box 861.
[62] Ambruster to Wallace, December 20, 1937, FDA File on Sulfanilamide-Massengill, Vol. 1.
[63] Carroll to Copeland, December 18, 1937, Petitions and Memorials File, 75A-F6.

clear in a speech to his colleagues. The Congress was in special session, he declared, because the President felt there were emergency matters requiring immediate attention. FDR had yet to present these matters to the Congress. Why not then, Rees declared, use the time to deal with S. 5: "We talk about emergency measures. This is a measure which can well come under this classification. If there ever was need for legislation on food and drugs for this country, that time is right now." Not only were the American people being duped continuously on matters of food and drugs, Rees continued, but up to this very moment seventy-three "innocent" people had died from the use of Elixir Sulfanilamide. "In view of recent experience we should give immediate attention to this important question."[64]

It was the function of Royal Copeland and his House colleague Virgil Chapman, however, to bring the elixir disaster in an official way to the floor of their respective chambers. On November 16 and 17 the two legislators pressed resolutions calling for a report to the Congress on the drug tragedy by the Department of Agriculture. The resolutions passed each house unanimously. The USDA report was presented to the Congress on November 26. In thirty-four pages of text and documents it laid bare the whole shocking story, from the failure of Massengill to test his elixir for toxicity to the technicality under which FDA was able to enter the case.[65] By the time of the report the women's organizations and other proponents of a new drug law were publicly emphasizing the fact that even if S. 5 had passed into law the sulfanilamide tragedy would still have taken place. S. 5 had no provisions to control new drugs entering the market.[66] Equally shocking, though less publicized, was the fact that the original 1933 bill, S. 1944, would have prevented the disaster. Miss Lamb ex-

[64] *Congressional Record*, 75th Cong., 2nd Sess. (November 24, 1937), 549-50.

[65] *Secretary's Report*, Senate Documents 124, 75th Cong., 2nd Sess.

[66] National League of Women Voters *News Letter*, October 28, 1937, FDA Scrapbooks, Vol. 16.

plained what had happened quite plainly in a letter to Alice Wright. "As you know," she wrote of S. 1944, "that bill was peddled up and down the halls of Congress without finding a taker—only advice—and as a result of that advice the licensing provision was reluctantly struck out."[67]

The elixir tragedy was undeniable evidence of the necessity of FDA control over new drugs. Even Clarence Lea, who had been less than helpful to the cause of a new food bill, was convinced. Several days after the USDA report he called the drug unit's Paul Dunbar to verify the fact that the present version of S. 5 would not prevent a recurrence of the elixir episode. Dunbar had to reply that the bill would not prevent a similar situation in the future. Lea then requested that FDA draft an appropriate amendment for the bill.[68] In fact, the process was already underway. Royal Copeland had made a similar request some time before USDA's sulfanilamide report.[69]

On December 1, 1937, the Senator introduced S. 3073.[70] This bill provided that manufacturers seeking to place new drugs on the market must first furnish the Secretary of Agriculture with records of their testing, a list of the drug's components, an explanation of manufacturing processes to be followed, and examples of labeling and such samples as were requested. The Secretary would then certify the drug as safe for sale or notify the producer of reasons for refusing certification. The bill was referred to committee. In the House, Virgil Chapman introduced a similar measure, H.R. 9341.[71] The provisions were essentially the same as S. 3073 though FDA came to favor the Chapman version, presumably because of its greater legal clarity.[72] Ultimately H.R.

---

[67] Lamb to Wright, November 16, 1937, Correspondence, Box 861.

[68] Dunbar, "Memories of Early Days of Federal Food and Drug Law Enforcement," *FDC Law Jnl.* 14 (February 1959), 137.

[69] Ole Salthe to Charles Crawford, November 24, 1937, Correspondence, Box 859.

[70] See Charles W. Dunn, *Federal Food, Drug, and Cosmetic Act*, 1018.

[71] *St. Louis Star-Times*, February 7, 1938, FDA Scrapbooks, Vol. 17.

[72] Salthe to Copeland, February 16, 1938, Petitions and Memorials File, 75A-F6.

9341 would be incorporated into S. 5 and pass into law.

Both bills brought an excited reaction from the drug industry, which had expected a licensing amendment. Vick Chemical Company's Richard K. Hines had made such a prediction to the press, as had *Drug Trade News*.[73] Still, portions of the industry were not happy with the actual event. *Proprietary Drugs* made that clear in its February issue. If these bills passed, the periodical lamented, "the Food and Drug Administration would become absolute dictator, the overlord of the drug industry." *Proprietary Drugs* called the Chapman bill in particular "an arbitrary and dictatorial proposal accomplishing no benefit to the public without crippling the supply of inexpensive and valuable preparations useful for self-medication."[74]

The trade generally agreed that some extension of FDA's power over new drugs was necessary, but they felt that the Copeland-Chapman method of regulation was too extreme. They favored weaker schemes such as those advocated by the Proprietary Association's James Hoge or trade lawyer Charles Dunn. Under Hoge's plan, manufacturers would submit to the Secretary materials similar to those called for in the current bills. The difference was that if the Secretary lodged no objection in a given period of time the product could be placed on the market. If the Secretary did have objections USDA would have to go to court for an injunction to halt the sale of the drug.[75] Dunn had an even weaker proposal. He called for a simple addition to the misbranding section of S. 5 making the failure to perform adequate tests on a drug subject to misbranding action. As Walter Campbell explained to Copeland, Dunn's suggestion meant nothing over what was already in S. 5. Dunn had lost sight

[73] *New York Times*, December 7, 1937, Correspondence, Box 863; Lamb to Wright, November 16, 1937, Correspondence, Box 861.

[74] "Chapman Bill H.R. 9341," *Proprietary Drugs* 24 (February 1938), 5ff.

[75] Lamb to Robert Littell, February 4, 1938, Correspondence, Carton 124, Acc. 52-A89, RG 88, FRC.

of the fact that penal rather than preventive action would not prevent another elixir disaster.[76]

Both Dunn and Hoge sought diligently to get concessions out of FDA and Copeland. Copeland talked with them, but that was all.[77] The trade fared a bit better in the House. The House version of S. 5 in which H.R. 9341 was incorporated did allow sales to proceed unless the manufacturer's application to the Secretary of Agriculture was rejected within sixty days.[78] This concession was not all that the trade hoped for. When the House version of S. 5 was reported in April 1938, *Proprietary Drugs* criticized bitterly the new drug clauses, deeming it unfair that FDA had the discretion to decide on the validity of a manufacturer's testing: "Let each manufacturer review his past experience with the FDA. . . . Can he recall an instance where the tests which he thought were adequate were so regarded by the officials?"[79]

Yet the trade did little more than grumble about the new drug provisions. The industry must surely have known that they were lucky to get the concession they did get. They were in a very poor position to make demands or draw hard battle lines on this particular matter. The wake of the Elixir Sulfanilamide affair was far too strong. That wake was easily apparent when Copeland brought S. 3073 to the Senate floor for a vote in May 1938. There was essentially no debate. Senator Vandenburg lamely brought to the attention of his colleagues the fact that objections had been raised to the bill. Copeland commented curtly that "there are objections, and there will be objections from now to the end of time, but, so far as I can judge, the bill is in such form that it is safe to pass it." The Senate agreed. Minutes later S. 3073 passed by a unanimous vote.[80]

[76] Campbell to Copeland, February 7, 1938, *ibid.*
[77] Salthe to Copeland, February 16, 1938; Copeland to Salthe, February 19, 1938, Petitions and Memorials File, 75A-F6.
[78] Dunn, *Federal Food, Drug, and Cosmetic Act,* 983.
[79] "Food and Drugs," *Proprietary Drugs* 24 (April 1938), 4.
[80] "S. 3073 Passes Senate," *ibid.,* May 1938, 9; *Congressional Record,* 75th Cong., 3rd Sess. (May 5, 1938), 8326.

A month before this event S. 5 had been reported to the House with the modified version of Chapman's H.R. 9341 incorporated. It was clear to all concerned that the nation would very soon have a new food and drug statute. The incorporation of controls over new drugs, backed as it was by the drama of Elixir Sulfanilamide, made the prospects for passage of S. 5 even more certain. This likelihood did not mean, however, that 1938 brought clear sailing for Senate bill 5. Indeed, even in the immediate wake of the elixir tragedy the Food and Drug Administration, and perhaps the public, sustained a significant defeat in March 1938. A second major battle over specific provisions of the bill was to come in the spring.

The former episode involved the old issue of advertising controls and got under way in January 1938 when S. 1077 came up for debate in the House. This was the Wheeler-Lea bill seeking to expand the powers of the Federal Trade Commission and, so far as Clarence Lea was concerned, designed to vest sole control over food, drug, and cosmetic advertising with the Commission. The minority membership of the House Commerce Committee, having earlier failed to halt this development, were determined to press their desires on the floor. The bill was accompanied by a vigorous minority report stating that FDA was better equipped to handle such advertising and that the cease and desist procedure of FTC was not an effective deterrent to fraudulent food, drug, and cosmetic claims.[81]

Debate on the measure was heated. The minority led by Carl Mapes and Kansas Representative Rees sought to strike those sections of the bill dealing with food, drugs, and cosmetics, but failed.[82] Next, and as an alternative, they sought to strengthen S. 1077 by providing for civil penalties up to a $5,000 fine in first-offense cases.[83] Clarence Lea skillfully defended the existing provisions of the bill. He lauded the

[81] Charles Dunn, *The Wheeler-Lea Act: A Statement of Its Legislative Record*, 195-97.

[82] Lamb to Littell, February 4, 1938, Correspondence, Carton 124, Acc. 52-A89, RG 88, FRC.

[83] Dunn, *Wheeler-Lea Act*, 195.

[171]

record of FTC and the effectiveness of the cease and desist method of control. At the crucial moment he announced that he had discussed the whole matter with Commissioner Ewin Davis of FTC. Davis had stated that the Commission did not want increased first-offense penalties, that such powers would be destructive to the successful operation of his agency. With the weight of the Commission thrown against the minority, the battle was lost. S. 1077 passed the House by an overwhelming vote on January 12.[84]

FDA was apoplectic over House passage of the bill. Ruth Lamb wrote a friend that S. 1077 "is itself really a piece of false advertising, since it purports to regulate advertising in the interest of consumers."[85] The Food and Drug Administration began a calculated effort to stop the Lea measure from becoming law, at least as it currently read. Miss Lamb began an extended correspondence to explain to all who would listen exactly how weak the bill was. To her, the worst aspect was that even the limited criminal penalties provided could be invoked only after proving fraudulent intent on the part of an advertiser. FDA had labored under the same handicap for years trying to enforce the Sherley amendment, and well understood the unfortunate limitations of such a provision.[86]

FDA's J. J. Durrett had for some time held a personal letter of introduction from Dr. Harvey Cushing to James Roosevelt. To this point, Durrett wrote Cushing at the end of January, he had felt it was inappropriate to use the letter in any attempt to press the desires of his agency on the President. Now he had determined to use it. The passage of S. 1077 made "it necessary that some action be taken

<hr/>

[84] Lamb to Clark Gavin, January 28, 1938, Correspondence, Carton 124, Acc. 52-A89, RG 88, FRC; David F. Cavers, "The Federal Food, Drug and Cosmetic Act of 1938," *Law and Contemporary Problems*, Winter 1939, 19.

[85] Lamb to Littell, February 4, 1938, Correspondence, Carton 124, Acc. 52-A89, RG 88, FRC.

[86] Lamb to Ruth Braem, March 15, 1938, *ibid*. A large body of Miss Lamb's correspondence on this matter may be located in Carton 125, *ibid*.

which has the possibility of results."[87] Durrett hoped for a veto. Henry Wallace also hoped for a Presidential rejection and he pressed this course on Roosevelt.[88] In this anger over the "fraud joker" FDA was joined by many consumer organizations. The women's organizations cried out against the provision, and Consumers Union called upon FDR in behalf of its 60,000 members to veto the measure when it reached his desk.[89]

By early March, however, even FDA was sure the bill would pass the Senate. On March 11, Charles Crawford wrote David Cavers that "there is no question but what the FTC act will become law. All that seems required is for Wheeler to catch his breath long enough to bring up the conference report between his diatribes against the Administration's tendency toward 'dictatorship.'"[90] On March 14, he "caught his breath" and the conference report came up for debate. "The consumers of this country are being raped," Royal Copeland said of the bill. "The housewives of America will be the victims of evils that will be felt by every home in our country."[91] Yet Copeland was well aware that any chance to shift controls over food, drug, and cosmetic advertising to FDA was gone.

On the floor of the Senate he told his colleagues: "I am the kind of chap who knows when he is licked. I know the House will never agree to turn over to the Food and Drug Administration the control of advertising. . . . I am not fooled about that. So, for the moment, I concede the point."[92] What the Senator did attempt was to strengthen S. 1077 by increased civil penalties. It was a losing fight. The

[87] Durrett to Cushing, January 25, 1938, Correspondence, Carton 124, *ibid.*

[88] Wallace to D. W. Bell, March 17, 1938, *ibid.*

[89] Lamb to Braem, March 15, 1938, *ibid.*; Consumers Union to Roosevelt, March 14, 1938, OF 100-Miscellaneous, Roosevelt Papers.

[90] Crawford to Cavers, March 11, 1938, Commissioners' File, Box 12.

[91] Press release, "Remarks of Senator Royal Copeland Relative to the Conference on S. 1077," Copeland Papers.

[92] Dunn, *Wheeler-Lea Act,* 344-47.

bill passed the upper chamber on March 14, 1938, and was signed into law March 21, 1938.[93]

The loss of advertising authority was a significant blow to the FDA. The only redeeming factor was that the loss had probably facilitated the early passage of a new food law. At least *Tide* and *Printers' Ink* thought so. Clarence Lea now had fulfilled his desire to place advertising with FTC. Having done so, he would bring S. 5 out of committee and to the floor for a vote.[94] Indeed, once the FTC measure was through the House there were stirrings on S. 5. In the first week of March the House subcommittee handling the bill agreed on that body's substitute for S. 5.[95] It was reported to the whole House on April 14. Yet even with this event the advent of a new food and drug law was still one battle away.

[93] *Ibid.*, 347.
[94] "Our Old Friend, S. 5," *PI* 182 (February 24, 1938), 94-95; *Tide*, March 1, 1938, FDA Scrapbooks, Vol. 17.
[95] "Substitute for S. 5," *PI* 182 (March 10, 1938), 17.

VIII

CODA: RALLY ROUND THE APPLE

It is a question for the House to decide whether it is going to follow the recommendations of the apple growers' association . . . or the Food and Drug Administration.

> CARL MAPES
> HOUSE OF REPRESENTATIVES
> MAY 31, 1938

And, for all its changes, it is now the Copeland law—becoming that just as the gates of eternity were swinging ajar to admit its sponsor.

> PRINTERS' INK
> JUNE 23, 1938

WHEN Royal Copeland introduced Senate bill 2000 in January 1934, Vice President John Nance Garner had a piece of advice for him. Garner told the New Yorker that such controversial legislation should be drawn and passed by both Houses, then really be written in the conference committee.[1] Perhaps this was not the way Copeland would have liked things to go, but Garner proved a good prophet. In the spring of 1937 Chairman Clarence Lea of the House Commerce Committee had announced that his group intended to draft their own version of S. 5. The Senate bill was still locked up in the House committee in January 1938 while Lea's Federal Trade Commission measure was on its way to the upper house.

In January, also, Copeland conferred with his House counterpart as to when S. 5 would be reported. Lea stated that the bill would be out in about two weeks. He advised Copeland further that the committee was seeking to cut

[1] Ole Salthe, "A Legislative Monument to Senator Royal Copeland," *FDC Law Jnl.* 2 (June 1947), 257.

out the controversial aspects of the measure so as to make it easy to pass. By this point at least Copeland had accepted Garner's admonition. The Senator wrote his legislative aid, Ole Salthe, "this seems to me a good idea because really, in the last analysis, we have to right [*sic*] it in conference anyway."[2] The matter proved to be not quite that simple. For one thing, Lea added to the bill one of the most controversial sections to that date. Secondly, Lea did not by any means get S. 5 out of committee in January.

The California Representative fully intended to hold back on S. 5 until he was assured that S. 1077 would become law and vest all advertising control powers with the FTC. Hearst's *Drug World* openly boasted of Lea's intrigues on S. 1077 as early as the spring of 1937.[3] Even after his pet measure became law Lea seemed in no hurry to deal with S. 5. By April 1938 Copeland was still trying to get some action in the House. "Dear Brother Lea," he wrote on April 11, "I feel like urging you again to push the matter, because of the shortness of the time. It would be a great pity to delay the matter until it's too late."[4] If S. 5 were not passed by the House soon, Copeland had doubts that he would be able to get any action in the Senate since the end of the session jam would soon be underway. This would mean that the whole effort for a new food-drug law would have to begin anew in January 1939.

Nor was Copeland the only one concerned that a new law would be delayed another year. Many of the affected industries were concerned. A chief reason was the longstanding threat of separate and different state drug statutes. The Proprietary Association's James Hoge reminded the trade of that menace again in the spring of 1938. Three states had now passed their own laws, he warned. Bills were pending in twelve more as well as Puerto Rico. A uni-

[2] Copeland to Salthe, January 8, 1938, Petitions and Memorials File, 75A-F6.

[3] Ruth Lamb to Helen Woodward, January 1, 1938, Correspondence, Carton 125, Acc. 52-A89, RG 88, FRC.

[4] Copeland to Lea, April 11, 1938, Petitions and Memorials File, 75A-F6.

form federal measure should be passed at once.[5] This demand for action was reinforced by the still potent wake the Elixir Sulfanilamide disaster had caused. Periodically journals such as *Commonweal* reminded their readers that there was still no public protection from recurrence of such a tragedy.[6]

Any new incident of abuse in the food-drug-cosmetic field immediately harkened the press back to Massengill's remedy and to the fact that the public lacked protection. In March a Kentucky woman was blinded by a poisonous eyelash dye. The *Memphis Commercial Appeal* linked it with memories of sulfanilamide: "Loath as we are ordinarily to raise the old cry of 'there ought to be a law,' this business obviously calls for . . . control."[7] In March also, six women died in Florida from injections of a so-called cancer cure sold in drugstores under the name "Ensol." The *Springfield* (Ill.) *Register* remembered the fatal elixir. So did other papers from McKeesport, Pennsylvania, to Charlotte, North Carolina. "What is needed, plainly," the *Charlotte News* stated in echoing the sentiments of its sister journals, "is a new and more stringent food and drug act."[8]

Whatever Clarence Lea's preference, such pressures dictated that he must get S. 5 out of committee before the close of the session. This was done at length on April 14, 1938. Lea described his version as containing "substantially all the features of the old law that have proved valuable," as well as amplifying and strengthening "the provisions designed to safeguard the public health."[9] There were those who vigorously disagreed at least in regard to one provision. It was bad enough that advertising had been withheld from FDA, but now there was Section 701 relat-

[5] Hoge, "Effects of Conflicting Food and Drug Legislation," *PI* 183 (May 19, 1938), 57, 60.
[6] T. Swann Harding, "How to Commit Legal Murder," *Commonweal* 28 (May 6, 1938), 40-42.
[7] March 10, 1938, FDA Scrapbooks, Vol. 17.
[8] *Charlotte News*, March 31, 1938; *Springfield* (Ill.) *Register*, April 7, 1938; *McKeesport News*, April 9, 1938; all in FDA Scrapbooks, Vol. 17.
[9] *Washington Post*, April 16, 1938, *ibid.*

ing to court review. This section provided that appeal suits might be entered to enjoin the Secretary of Agriculture from enforcing new regulations which USDA had promulgated. These suits could be entered in any of the over eighty federal district courts. The courts would have power under prescribed circumstances to take new evidence bearing on the regulations and to then determine the validity of those regulations. An adverse decision for the government by any single court would at least prevent enforcement for months if not years.[10]

This cumbersome procedure of appeals on identity, quality, and labeling regulations by the Secretary immediately aroused the anger of the Women's Joint Congressional Committee, the coordinating body for fourteen national women's organizations.[11] Secretary of Agriculture Henry Wallace reacted by stating that the department considered it better to continue the old law in effect than to have S. 5 enacted with the review provision. A vigorous minority report was also registered by dissenters within the Commerce Committee. The minority called the appeal procedure "unprecedented" with respect to regulation machinery by federal agencies, and warned that it would hamstring the control efforts of FDA.[12] Rumor had it that the court provisions had never been brought up for discussion in the full committee. Lea, the story ran, had written the amendment himself without informing his colleagues. He had then rushed it through an executive session of the committee to become part of the House version, catching dissenters totally off guard.[13] Certainly minority spokesman Virgil Chapman was shocked. He called the provision "an extraordinary extension of jurisdiction . . . never heretofore seriously

[10] David F. Cavers, "The Food, Drug, and Cosmetic Act of 1938," *Law and Contemporary Problems* 6 (Winter 1939), 21.
[11] "The Food, Drug, and Cosmetic Act," *Jnl. of Home Economics* 30 (June 1938), 401.
[12] *Ibid.*; *Congressional Record*, 75th Cong., 3rd Sess. (April 21, 1938), 7415.
[13] *Louisville Times*, April 27, 1938, FDA Scrapbooks, Vol. 17.

advanced in the entire history of the country."[14] Battle lines began to form.

The court review issue was the last major fight in the long history of efforts to pass what became the 1938 Food, Drug, and Cosmetic Act. The proposal to give the Secretary the power to make regulations carrying the force of law had been a part of every food-drug bill since the original 1933 version. The power was involved in a number of provisions of the respective bills, but one single provision was chiefly responsible for the battle as it developed in the spring of 1938. This was the section which defined as adulterated any food containing poisonous ingredients, subject, however, to the power of the Secretary to establish tolerance levels where poisons could not be wholly eliminated. In this matter the most seriously affected foods were fresh fruits and vegetables where efforts to remove poisonous insecticide spray residue had never been completely successful.[15]

FDA had long devoted a portion of its appropriations to assuring that spray residues did not exceed tolerance levels set by the Secretary. When specific actions by the agency were contested in the courts, however, it had been necessary for the drug unit to show not only that a producer had exceeded the tolerance level but also that the amount of poison contained rendered the product dangerous to health. To prove before a jury that a small amount of poison residue was actually dangerous to health was a very difficult task. If the Secretary were given the power to promulgate regulations carrying the force of law as proposed in the successive food-drug reform bills, the problems of FDA would be greatly lessened. The agency would simply have to show that tolerance levels had been exceeded. This having been established the only recourse left to a defendant would be to attack the regulations as unconstitutional. Hence the importance of the court review amendment.[16]

[14] *New York News*, April 25, 1938, *ibid.*
[15] Cavers, "Food, Drug, and Cosmetic Act," 15.
[16] *Ibid.*

If the Lea provision stuck, food producers hit by tolerance regulations would not be quite as free as under the
1906 statute, but they could stall the inception of regulations almost indefinitely by an endless round of appeals.
Moreover, if the aggrieved could prove to the courts that
the introduction of new evidence was warranted as prescribed in the clause they could create a situation exactly
like that which prevailed under the old law. FDA would
have to justify in court the tolerance level established. The
burden of the protection of the food and drug producers'
constitutional liberties was largely carried in the spring of
1938 by the International Apple Association.

According to David Cavers, the genesis of the Association's concern on this matter went back to a severe shock
they had received in the spring of 1933. When the new
Assistant Secretary, Rexford Tugwell, became aware of the
intricacies of tolerance levels questions, he had moved to
reduce the level on lead arsenate residue from 0.02 grain
per pound to 0.014 grain. The resulting outburst of protest
from apple producers was so strong that Secretary Wallace
was forced to return the tolerance level to 0.02, but the
experience had a lasting impression on the industry. They
began to realize the dangers inherent in providing the Secretary with the power to issue regulations carrying the force
of law. Their objective became to broaden the courts' power in passing on regulations set by USDA. If the judgment
of district courts on such questions could be substituted
for the opinion of experts in the Agriculture Department,
as David Cavers put it, apple producers "could face the
future with aplomb."[17]

These producers first began to press their case at the
1935 Senate hearings on S. 5. Speaking for the International
Apple Association, Samuel Fraser had protested current
clauses in the bill. He argued that if S. 5 were passed, the
burden of proof on such matters as tolerance levels would
be shifted to the defendant, a procedure out of accord
with the American legal tradition. "Tom Jones," he insisted,

[17] *Ibid.*

would be unable any longer to question the facts surrounding the decision which proclaimed him guilty. Fraser had "certain amendments" to offer which would safeguard the civil liberties of the producer. They had to do with court review procedure.[18] Alas for the apple producers, their grievances and case were largely lost amidst the currently more impressive issues of multiple seizure regulations and the question of the location of advertising control authority. By 1938 these issues were gone and with the help of Clarence Lea, whose constituency included many fruit growers, the appropriate time had come to press the cause.

The court review sections of S. 5 aroused in 1938 more protest from consumer-oriented sources than even the Bailey amendment had.[19] The women's organizations took the lead. They were not merely angry, they were enraged. They threatened to oppose the bill altogether unless drastic alterations were made in Section 701.[20] They charged that the appeal clause had "devitalized" the entire measure.[21] The women began a large-scale campaign to bombard their elected representatives with letters of protest.[22] The reaction of organized consumer groups was only slightly more militant than the reaction of the Food and Drug Administration itself. Charles Crawford was extremely upset by the turn of events. In a letter to David Cavers he castigated "the apple crowd." He charged that Lea's move was made purely to fill "a political need in the forthcoming campaign in which the support of the fruit growers will be a substantial element in his success at the polls."[23]

As previously mentioned, Henry Wallace, speaking for USDA, publicly announced his preference for retaining the

[18] *Senate Hearings on S. 5* (1935), testimony by Fraser, 265-69.
[19] Cavers, "Food, Drug, and Cosmetic Act," 21.
[20] *St. Louis Post-Dispatch*, April 28, 1938, Correspondence, Carton 124, Acc. 52-A89, RG 88, FRC.
[21] *Philadelphia Record*, April 29, 1938, FDA Scrapbooks, Vol. 17.
[22] Lamb to Salthe, June 17, 1938, Correspondence, Carton 125, Acc. 52-A89, RG 88, FRC.
[23] Crawford to Cavers, March 23, 1938, Commissioners' File, Box 12.

1906 law rather than accepting Lea's court review clauses. He went further. In April the Secretary wrote the President, labeling Section 701 as "without a parallel in Federal legislation." He informed the President that he would have no alternative but to recommend veto if the House version of S. 5 were passed in its present form.[24] Roosevelt was apparently quite disturbed by this reaction. He sought the opinion of the Attorney General on the matter of review. The Attorney General promptly replied that his office was "in accord with the observations made by the Secretary in condemning the provisions." The procedure was so cumbersome as to make efficient administration of the bill virtually impossible.[25] USDA continued during May to press on FDR its case for adamant resistance.[26]

It would appear, however, that Copeland's office took a view somewhat different from that of the Agriculture Department. The Senator's legislative assistant Ole Salthe, at least, was more than a bit irritated at the actions of USDA. In late April he wrote Copeland complaining that the department seems to "insist upon making our job that much harder." He differed with the view of the Secretary, but, more important, he felt compromise was possible without, as the department had done, "putting a blight on the whole bill." Salthe was also angry at Ruth Lamb. "No wonder the proprietary people fight so hard," Salthe grumbled. They had recently been very aggrieved over the dangerous drugs warning clause of the bill and now Miss Lamb had given them new ammunition. In a recent letter she had commented so critically on common proprietary remedies as to "put the Department on record against self-medication."[27]

On April 29 Salthe again aired his grievances about

[24] Wallace to Roosevelt, April 5, 1938, OF 375, Roosevelt Papers; Wallace to Roosevelt also located in Carton 125, Acc. 52-A89, RG 88, FRC.

[25] Homer Cummings to Roosevelt, April 11, 1938, OF 375, Roosevelt Papers.

[26] USDA to Roosevelt, May 17, 1938, *ibid.*

[27] Salthe to Copeland, April 22, 1938, Petitions and Memorials File, 75A-F6.

USDA to his chief. The food-drug bill was in serious trouble. A major factor in the trouble was the House Committee minority report which Salthe felt was "written largely by the Food and Drug Administration." The strong opposition statement of the Secretary was also to blame. Differences could have been ironed out without all this bravado. Now there was a very good chance that S. 5 would not come up for a vote in the House at all. FDA-oriented criticism had so embittered Clarence Lea as to make him "more than ever opposed" to any suggestions which might come from that agency. Moreover, continued Salthe, Speaker Rayburn was no longer enthusiastic about bringing the measure up. S. 5 had become controversial, and the Speaker was reluctant to see any controversial measure reported out for action so late in the session. Salthe urged Copeland to go directly to Rayburn and Lea in the interest of the bill.[28]

The Senator did plead his case with his two House colleagues.[29] By the latter part of May he was also besieging Franklin Roosevelt to lend some help. "It is really too bad that insistence upon an amendment proposed at the last moment should defeat such much needed legislation," he wrote FDR on May 23. Could the President use his influence to get the bill into conference?[30] Eventually Roosevelt would take a hand but not at this point. The matter of bringing S. 5 up for a vote in the House was prompted by other considerations. One factor was very likely the widespread expressions of discontent which members of Congress found in mail from constituents.[31]

The discontent owed much to the fact that the sulfanilamide disaster remained fresh in the public mind. On into May periodic reminders that nothing had been done to assure that a similar situation would not occur appeared in the press. Some of these journals were very limited in cir-

[28] Salthe to Copeland, April 29, 1938, *ibid.*
[29] Salthe to Charles Crawford, May 4, 1938, Commissioners' File, Box 12.
[30] Copeland to Roosevelt, May 23, 1938, OF 375, Roosevelt Papers.
[31] *Ibid.*

culation such as the *Great Falls* (Mont.) *Tribune*. "Months have passed . . . yet Congress fails to give us an effective pure drug law," that paper editorialized in May. "This is inexcusable."[32] Some of the journals had a much wider audience. In May *Survey Graphic* reprinted the poignant plea for action to FDR from Mrs. Maise Nidiffer whose young daughter had died from the use of the elixir. Mrs. Nidiffer's dramatic description of the loss had even become a part of the Secretary's report to Congress on the drug disaster. *Survey Graphic* wanted to make sure the public did not forget that "little body tossing to and fro"—"that little voice screaming with pain," as Mrs. Nidiffer related her child's last days. Must we have another series of tragic deaths, the journal asked, before action on a new law is taken?[33]

All of these periodic reminders of inadequate regulation in the drug market surely played some part in moving S. 5 to a House vote. The specific commotion over the review section itself certainly played a major role, at least for Clarence Lea. The whole matter of the food-drug bill was becoming a political hot potato. By early May, Lea was anxious for compromise, according to FDA's Charles Crawford, in order "to nullify the ammunition, which might be very effectively used against him." Even the Department of Justice had informed the California Congressman that his amendment was impractical and had urged him to accept some compromise.[34] Adamant defense of the apple interests was becoming too costly.

In the meantime, influenced by rumors that Lea might accept some change, Copeland's office along with FDA worked to draft a suitable compromise. The idea was to have it ready, as Ole Salthe put it, to present in conference committee "if the bill ever passes the House."[35] The FDA compromise plan was to allow appeals from the regulations

[32] May 14, 1938, Correspondence, Carton 124, Acc. 52-A89, RG 88, FRC.
[33] In FDA Scrapbooks, Vol. 18.
[34] Crawford to Salthe, May 7, 1938, Commissioners' File, Box 12.
[35] Salthe to Crawford, May 4, 1938, *ibid.*

of the Secretary but only in the District Court of Washington, D.C., rather than all across the country. Ole Salthe liked the FDA proposal. His concern was that Virgil Chapman and Carl Mapes might resist any compromise. "It would be fatal if they continue the controversy," he wrote Crawford. He asked that Crawford use his influence with the two representatives to get their assistance in bringing the bill to conference.[36]

Salthe's letter to Crawford was dated May 10. Five days earlier, Copeland had forced Senate passage of S. 3073, his Sulfanilamide bill. Salthe was not sure why the Senator had pressed the matter. He thought the aim might have been a bit of further strategy to force action in the House.[37] Very likely Salthe was correct and, even more likely, the strategy was good. Lea was going to have to make some move in the immediate future. The pressures behind him were building rapidly. Indeed by May 10 Salthe had heard through James Hoge of the Proprietary Association that the Californian planned to call a full meeting of the Commerce Committee around May 17 with an eye to getting S. 5 to a vote.[38]

The bill was brought up on May 31, 1938, and debated on that day as well as the next. While the debate touched many areas of the measure, as was expected the major issue was the court review provision, which remained essentially unaltered. The opposition to Section 701 was led by Representatives Carl Mapes, John Coffee, and Charles Wolverton. There was little new in the attack. The House was told that the court procedure was too cumbersome and that new regulations by the Secretary would be tied up in courts for extended periods of time before they could go into effect. It was charged again and again that the appeal provisions were designed to shield manufacturers of patent medicines, cosmetics, and foods at the expense of the public welfare. The apple growers were chiefly responsible for the

[36] Salthe to Crawford, May 10, 1938, *ibid.*
[37] *Ibid.*
[38] *Ibid.*

current deadlock, but, of course, court review would apply to all regulations promulgated by the Secretary.

The dissenters insisted that the review clause was unnecessary anyway since any citizen who felt his rights had been violated by a decision of the Secretary could under legal procedure go to court and ask that the Secretary be enjoined from enforcing his decisions. The opponents of Lea's amendment repeatedly pointed out that USDA and even the Justice Department strongly disapproved of the review section. Several attempts were made to amend the objectional clause and at length to recommit the whole bill to committee. Each effort failed. Clarence Lea and other proponents of the court review section answered their antagonists in able fashion, basing their case on the need to put some checks on the broad grant of power which would be handed to USDA when and if the bill was passed.

In fact, the issue on which House members must determine their vote had been quite succinctly drawn in the early minutes of debate on May 31. "It is a question for the House," as Carl Mapes drew the negative position, "to decide whether it is going to follow the recommendations of the apple growers' association in writing the section or the recommendations of the Food and Drug Administration."[39] Clarence Lea at least as aptly stated the affirmative posture. "The practical problem presented by court review," he told his colleagues, "is whether you are in favor of a government by edict or whether you favor a government by orderly procedure, a government under which the citizens shall have a right to be heard."[40] In the end Lea's position was all-persuasive. After a motion to recommit S. 5 to committee failed, the food bill passed the House. As expected, the Senate disagreed on the lower chamber version and requested a conference.

In reporting this event the press took notice that the

[39] House debate on S. 5 may be found in *Congressional Record*, 75th Cong., 3rd Sess. (May 31 and June 1, 1938), 10,221-10,222; 10,226-10,254; 10,316-10,330. For convenience, the quote may be found in Dunn, *Federal Food, Drug, and Cosmetic Act*, 843.
[40] Dunn, *ibid.*, 851.

House had overridden the objections of Secretary Wallace on the court issue. Such a step was newsworthy.[41] For consumer organizations the event was a public outrage. The national women's organizations let it be known again that they preferred to lose the entire bill rather than accept the review section. They would urge FDR to veto S. 5 if the odious court provision were not eliminated.[42] The House judicial clause, according to the June *News Letter* of the National League of Women Voters had "defeated consumer protection."[43] In urging Senator Josiah Bailey and other Congressmen to eliminate the "joker clause" the National Consumers League, like the women, charged that Section 701 "completely vitiates any advance that this measure might otherwise offer."[44] Consumers Union advised its readers that the "skilled lawyers of the International Apple Association" had finally hit upon a way to successfully stymie the enforcement efforts of FDA. These lawyers, CU continued, "will receive the undying gratitude of all those patent medicine and food manufacturers who for the past five years have opposed the enactment of decent food and drug legislation."[45]

In the immediate period after House passage many of the discontented began to deluge FDR with calls for Presidential action. Among the first to do so were the national women's organizations. Their plea, through Miss Marguerite Wells, President of the National League of Women Voters, included also an interesting commentary on where the real source of White House concern with passage of a new statute was located. Miss Wells wrote the President that at various times the women had conferred with Mrs. Roosevelt about the food bill and that she had made her-

[41] See, for example, *New Orleans Times-Picayune*, June 2, 1938, FDA Scrapbooks, Vol. 18.

[42] Lamb to Mr. P. Fuller, June 6, 1938, Correspondence, Carton 125, Acc. 52-A89, RG 88, FRC.

[43] In FDA Scrapbooks, Vol. 18.

[44] General Secretary, National Consumers' League to Bailey, June 6, 1938, Bailey Papers.

[45] Consumers Union *Reports*, June 1938, FDA Scrapbooks, Vol. 18.

self personally responsible for calling such communications to FDR's attention. Eleanor had suggested that on occasions when she was absent from the capitol the women might communicate directly with the President. Presumably this was such an occasion. Miss Wells enclosed a statement by fourteen national women's organizations asking that action be taken to eliminate the court section from S. 5.[46]

Large numbers of letters from private parties asking intervention arrived at the White House. The bulk of these were probably sent as a result of a call to action made by Consumers Union to its membership. What CU wanted, however, was for FDR to veto Copeland's bill, not merely force compromise on the court provisions. Consumers Union felt the whole bill was now too weak. CU-sponsored demands for veto continued to arrive long after a court compromise had been arranged. This militant organization urged the President to get behind a CU-sponsored drug bill recently introduced in the Congress by Representative John Coffee of Washington.[47]

USDA continued its militant resistance to the bill as passed by the House. On June 9 Walter Campbell made FDA's position quite clear to Congressman A. J. Sabath of Illinois. If changes were not forthcoming in conference, the drug chief wrote, ". . . it is the deliberate judgement of the Department . . . that the public would fare better by a continuation of our present operations under the present antiquated and inadequate statute."[48] About the same time Secretary Wallace again apprised the President of his opposition to the House version. The Secretary pointed out that most of the conferees felt there was a danger of veto if the court section were not altered. The likelihood of a suitable report very much depended on the

[46] Wells to Roosevelt, June 6, 1938, OF 375, Roosevelt Papers.
[47] Josephine Kane to Roosevelt, June 23, 1938, Correspondence, Carton 124, Acc. 52-A89, RG 88, FRC. Many such letters may be located in the above file as well as in OF 375, Roosevelt Papers.
[48] Campbell to Sabath, June 9, 1938, Correspondence, Carton 125, Acc. 52-A89, RG 88, FRC.

President's "emphatic indication" not to compromise. He urged the Chief Executive to "hold firm."[49]

At this crucial hour the President did "hold firm." He gave no indication that he would accept S. 5 with the unsatisfactory House provisions included. According to the *St. Louis Post-Dispatch* he went a good deal further. That journal reported that FDR had specifically informed members of the House Commerce Committee that veto would be forthcoming if the bill reached his desk with the House court review clause still a part of the measure.[50] Whether the intervention was positive or simply negative in the sense of White House silence, it must be credited as a major factor in bringing about the court compromise embodied in the conference report dated June 11.

This arrangement provided that in cases of disagreement with regulations promulgated by the Secretary the aggrieved party might file an appeal with any of the ten federal circuit courts. While the agreement still made it possible for the courts to enjoin the Secretary from enforcement of a regulation temporarily or permanently, it limited the circumstances under which new evidence could be introduced to the court as well as the circumstances under which the court might order rehearings. The greatest advantage of the compromise, however, was to reduce the points of appeal from the eighty-five district courts to the ten circuit courts.[51]

Nor was the court review change the only significant compromise arrived at by the conference committee. Charles Crawford had feared that the committee would combine the worst features of both the House and Senate bills. In fact, the reverse generally proved true. For this development a good deal of credit apparently belonged to Ole Salthe. Copeland brought Salthe to the committee meeting and Salthe came bearing a number of proposed

[49] Wallace to Roosevelt, June 8, 1938, *ibid.*
[50] June 12, 1938, FDA Scrapbooks, Vol. 18.
[51] "Food and Drug Bill Passed at Last," *Business Week*, June 18, 1938, 37. See also Dunn, *Federal Food, Drug, and Cosmetic Act*, 987.

changes which FDA felt should be made in the bills of both chambers. According to Crawford, writing some years later, Salthe found the conference in a mood to compromise. He took the occasion not only to get the stronger provisions between the bills of the two chambers into the final version but also at points to add stronger language than had been in either measure.[52]

While Crawford did not spell out all such changes the conference version of S. 5 did show a number of points of strength along lines suggested by him. Typical examples were in the sections on label disclosure of specified ingredients in food and drug products, variation in drug products from standards of official compendiums, and the clause on multiple seizure powers. On the matter of ingredient disclosure the conference accepted the stronger Senate provision which did not allow manufacturers the option to escape label requirements by filing their formula with the Secretary as provided in the House measure. In the case of variation from official standards the stronger House section was retained. This section forced manufacturers to specify exact variation from official standards on labels rather than allowing a simple declaration of the drug's strength, quality, and purity as provided in the Senate bill. On seizure the measures of both houses essentially allowed multiple actions in misbranding cases where the Secretary deemed that an article was "dangerous to health" or that the labeling was "in a material respect, false or fraudulent." The conference version contained a stronger language. In that version multiple action would be allowed when the Secretary deemed "without hearings by him or any officer or employee of the Department" that an article was dangerous to health or that labeling was "fraudulent, or would be in a material respect misleading."[53]

[52] Crawford to Milton Handler, June 1938, Correspondence, Carton 125, Acc. 52-A89, RG 88, FRC; Crawford, *F&D Review* 36 (October 1952), 201.

[53] Comparison of provisions among the Senate, House, and conference versions of S. 5 may be made with convenience in Dunn, *Federal Food, Drug, and Cosmetic Act.* Dunn reproduces all three

Both chambers promptly ratified the conference report, and S. 5 was presented to Franklin Roosevelt on June 15. In general, governmental proponents of a new food-drug bill were pleased at the compromise. In the House, Virgil Chapman expressed his satisfaction: ". . . let me say that if the bill reported to the House originally had contained a court review section such as that contained in the bill we are adopting today, I am sure the gentleman from Michigan and I never would have filed that minority report."[54] In the Senate, Copeland told his colleagues that "we now have a bill of which we may be proud."[55] FDA was equally pleased, indeed, as Charles Crawford put it, "greatly and pleasantly surprised." ". . . we seriously doubted," he wrote David Cavers, "that Senator Copeland would be able to swing the conferees to the acceptance of anything that would have been worthy of other than veto action."[56] "We think that the bill as finally agreed upon is a surprisingly good measure," Crawford further wrote to the Chief of FDA's Eastern District.[57]

The Food and Drug Administration had reason to be pleased. The bill did represent a vast improvement over the old 1906 statute. Yet the victory, particularly in the matter of court review, was not without important limits. The appeal procedure was a legal innovation and a restrictive innovation. In a curious way, according to law professor Ralph Fuchs, the Federal Trade Commission had come back to haunt the operations of FDA once more. The

versions. Specific contrasts as made above are as follows: disclosure of ingredient clauses, 783 and 786, 802 and 805, 978 and 982; multiple seizure power, 780, 798, 974; variation clause, 786, 804, 981. It should be mentioned that the conference-drawn variation clause was in one aspect weaker than the House version. The House provision allowed variations from official compendiums in strength but not in quality or purity. The conference provision allowed variation in all three.

[54] *Congressional Record*, 75th Cong., 3rd Sess. (June 13, 1938), 13,272.

[55] *Ibid.* (June 10, 1938), 11,543.

[56] Crawford to Cavers, June 14, 1938, Correspondence, Carton 125, Acc. 52-A89, RG 88, FRC.

[57] Crawford to W.R.M. Wharton, *ibid.*

ideas of the review section stemmed mainly from the provisions of the Federal Trade Commission Act. This act allowed for court review of FTC's cease and desist orders both in enforcement proceedings brought by the Commission and in injunction suits filed by parties to whom such orders apply. The vesting of jurisdiction in the circuit courts was also part of the FTC act.[58]

Contrary to the apparent view embodied in the court sections of the bill, however, FDA and FTC were not the same type agencies. The Commission was a quasi-judicial body and thus should be bound by more elaborate court review. FDA was strictly a regulatory agency, and the exercise of its rule-making duties required a fair degree of leeway. As early as the introduction of S. 2800 in 1934, Copeland argued that the delegation to the Secretary of rule-making power had been cut to the minimum. He contended that the legal discretion of the Secretary was confined to matters so complex and so changeable in the light of scientific progress that adequate protection of the public could not otherwise be offered.[59] Under the appeal procedure in S. 5, however, all the Secretarys' regulations regarding foods, drugs, and cosmetics were potentially subject to the verdict of the court. This procedure could be cumbersome and retard quick, effective regulatory action by the agency.

FDA was not singled out among regulatory agencies to suffer the additional legal burden. As Ralph Fuchs has explained, the review and, for that matter, the procedural provisions in S. 5 were part of a recent legislative tendency to impose more requirements upon regulatory bodies in the exercise of their rule-making powers.[60] Similar procedural and review provisions were to be found in the Fair Labor Standards Act of 1938, in the Bituminous Coal Act of 1937,

[58] Fuchs, "The Formulation and Review of Regulations under the Food, Drug, and Cosmetic Act," *Law and Contemporary Problems* 6 (Winter 1939), 49.

[59] Speech, "Senator Copeland's Statement about New Food and Drug Bill," Winter 1934, Copeland Papers.

[60] Fuchs, "Formulation and Review," 43.

and in the organic statutes of the Securities Exchange Commission, the Federal Communications Commission, as well as the Civil Aeronautics Authority.[61] Yet again, as Fuchs points out, what would become the 1938 Food, Drug, and Cosmetic Act did embody the most drastic consequences of this legislative tendency. The procedure set up resembled machinery used in dealing with such matters as utility rates. Employed in governing regulations on health and safety it was unique. Moreover, the formula as used in the past had generally applied the review procedure where decisions and orders were addressed to definite parties as contrasted with regulations of general application.[62] Thus FDA would labor under legal restrictions not characteristic of governmental agencies of like type.

There were other drawbacks and weaknesses about the Copeland bill. Certainly one weakness was the matter of seizure of misbranded goods. Perhaps the 1935 Bailey-Copeland compromise on multiple seizure had been an equitable one, given the sharp difference of views on that item. Yet, equitable or not, FDA would be more restricted in its seizure power once the bill became law than it had been under the 1906 statute. The failure to provide the Food and Drug Administration with advertising control powers was another weakness of the bill. The consumer would be better protected than ever before, in that specific power over false and misleading advertising had been vested with the Federal Trade Commission under the Wheeler-Lea act. It does seem true, however, that FDA would have been far more able to fulfill its charge as guardian over the operation of the food, drug, and cosmetic market if the agency had been empowered to regulate advertising.

Even in such matters as control over drugs entering the consumer market Copeland's bill would not prove a definitive panacea. This was made plain in April 1938 before S. 5 became law. Deaths in Florida from the use of a cancer

61 *Ibid.*, 46-47, 49.
62 *Ibid.*, 43, 49.

serum known as Ensol had helped the progress of S. 5 by
reminding the public of the sulfanilamide tragedy and that
little had been done to halt a recurrence. Yet, ironically,
as Ruth Lamb wrote Robert Littell of *Reader's Digest*, even
if S. 5 were law, FDA would not have had the power to
enter the Florida case. Copeland's measure contained a
provision specifically denying the drug unit authority over
products subject to the Virus, Serums, and Toxin Act which
dealt with authority vested in the Public Health Service.
Even worse on this occasion, because of a "loophole" in
the Toxin Act, PHS could not order seizure of outstanding
shipments of Ensol.[63] Moreover, it is pertinent to emphasize
here too that requirements in S. 5 on manufacturers to certify the safety of a drug applied only to new drugs and not
to drugs already on the market. Also, manufacturers seeking to introduce new drugs into the market were required
to provide evidence of safety alone. They were not compelled to prove that their products had therapeutic value
as such.

After the bill became law it would take time to get it into
operation. The scheduled effective date for full operation
was June 25, 1939. Long before that date FDA became
involved in an almost continuous battle with the affected
trades over the actual regulatory implementation of the
law.[64] Before June, manufacturers had gained a year's re-
prieve on portions of the labeling demands made by the
new law. Producers argued that the June date was an un-
due burden, and Congress had authorized the time extension for manufacturers who filed an affidavit of their problems with the food-drug unit. FDA was unhappy about the
concession. As one official commented caustically, "We'll
at least get a complete list of all food, drug and cosmetic
manufacturers."[65]

[63] Lamb to Littell, April 5, 1938, Correspondence, Carton 124,
Acc. 52-A89, RG 88, FRC.
[64] "Protest New Drug Regulations," *Business Week*, November 5,
1938, 36.
[65] "Postpone New Food, Drug Labels," *ibid.*, May 13, 1939, 46.

Yet with all the weak spots in the Copeland bill and the problems of getting it into full operation, S. 5 was a very good measure. The final product was well worth the five-year fight in terms of new protection offered to the American consumer. To begin with, all cosmetics except soap would be brought under the general control of the Food and Drug Administration. The new law would specifically outlaw cosmetics injurious to health. It would also prohibit traffic in foodstuffs which were dangerous to health. The 1906 statute had prohibited injurious food only when a poisonous substance was added. The Secretary was further allowed to set up emergency-permit control of food that might be injurious because of contamination with microorganisms. Traffic was forbidden in confectionery containing metallic trinkets and other inedible substances.

The new act would provide for government promulgation of a definition and standard of identity for foods as well as a reasonable standard of quality and fill for food containers. The old law contained no authority to establish definitions and identity standards. The authority to set up standards of quality and fill was limited to canned foods. Labeling on food for which no definition and standard of identity were fixed must, with minor exceptions, disclose ingredients by name.

Extensive improvement in federal control over the drug market came when S. 5 passed into law. Drugs used in the diagnosis of disease as well as any drug which affected the structure or any function of the body were brought under legal control. Likewise, all therapeutic devices became subject to FDA regulation, and devices must meet the general requirements set up for drugs. Certainly the most significant single feature of the law was the new-drugs provision. New drugs—devices were not included in this clause—could not enter the consumer market until they had been adequately tested to show that they were safe under conditions of use prescribed in their labeling. This innovation in consumer protection alone would have justified the five-year fight for a new law. Nor was the greater protection given the public the only consequence of the new-drugs clause. As James Harvey Young has explained, that provision of the law, by requiring pharmaceutical manufacturers to vastly expand their testing and research potential, proved an accelerating force in the discovery of new drugs.[66]

The new statute would require further that drugs recognized in official compendiums, now including the Homeopathic Pharmacopeia, must reveal any differences of strength, quality, or purity from those prescribed standards. The Wiley law required merely that labels bear a true statement of strength, quality, and purity; indication of specific variations from official standards was not necessary. The Copeland bill declared that nonofficial drugs were illegal unless their labels listed all active ingredients. They were also subject to legal action if actual standard of strength differed from the standards claimed. The 1906 statute made illegal only those drugs which *fell below* the strength claimed. Finally, FDA was no longer saddled with the necessity forced by court interpretation of the old law to prove that false label claims of curative effect were made with willful intent to deceive.[67]

On June 25 Franklin Roosevelt signed Copeland's food bill into law. The five-year struggle had come to an end. Both sides could lay down their arms and did so with relative satisfaction at the result. The affected industries still had some qualms but not enough to seriously regret the passage of the new law. Speaking for the advertising industry, *Printers' Ink* congratulated the New York Senator

[66] James Harvey Young, "Social History of American Drug Legislation," in Paul Talalay, ed., *Drugs in Our Society* (Baltimore, 1964), 227.

[67] Good brief references on the highlights of the new law are the following: "Highlights of New Food Law," *Food Industries*, August 1938, in Correspondence, Carton 124, Acc. 52-A89, RG 88, FRC; Copeland, "New Food, Drug, and Cosmetic Act," *Am. Perfumer*, July 1938, Copeland Papers; USDA, "Digest of the New Federal Food, Drug, and Cosmetic Act," June 27, 1938, Carton 124, Acc. 52-A89, RG 88, FRC. For the full text of the 1938 drug law see *United States Statutes at Large*, Vol. 52, 75th Cong., 3rd Sess. (1938), 1040-59.

on a job well done. "Largely through his wise generalship," that journal noted, the new law had been developed "from the impossible Tugwell Bill into a fairly satisfactory measure."[68] *Food Industries* felt that the food trade should have been given a separate law and partly blamed the other affected trades for the fact that they did not get it. Yet the food periodical also concluded that the new statute would bring some definite benefits to the industry. The statute would "only handicap materially those known to the industry as 'chiselers.'" On the whole *Food Industries* agreed with *PI*, "it is a satisfactory law."[69]

The drug trade was a bit more disturbed. *Proprietary Drugs* warned that the statute "is not to be regarded as simply a new food and drug law extending the scope of regulation. It is a law based upon an entirely new conception of regulatory control." It provided the government with "absolute dictatorial power" over the industry.[70] Yet, if *Proprietary Drugs* accurately reflected views in that trade, so too did the *Oil, Paint and Drug Reporter*. "Whatever the new obligations that the revised law may lay upon the industries affected," that journal stated, "to meet these will be far less disturbing than would have been further prolonged uncertainty of what was to be."[71]

Among proponents of strengthened food and drug legislation there was much satisfaction with the new act. There were, however, exceptions to this rule. Consumers Union fought to the end for a Presidential veto.[72] Rexford Tugwell spoke of the law as "a discredit to everyone concerned in it."[73] Such militant reformers felt ever so keenly the many compromises embodied in the legislation—compromises which meant the government would not be able to

[68] "Cure Thyself," *PI* 183 (June 22, 1938), 80.
[69] "New Food Law, at Last," *Food Industries* 10 (August 1938), 420; "A Double-Edged Sword," *ibid.*, October 1938, 572.
[70] Editorial, *Proprietary Drugs* 25 (June 1938), 1.
[71] *OP&D Reporter*, June 20, 1938, FDA Scrapbooks, Vol. 18.
[72] See Josephine Kane to Roosevelt, June 23, 1938, and sundry other similar letters in Correspondence, Carton 124, Acc. 52-A89, RG 88, FRC.
[73] Tugwell, *The Democratic Roosevelt* (New York, 1957), 464.

provide the minimum degree of consumer protection these idealists felt necessary. The harmful compromises were quite real and elements of this fuller view of consumer protection would in the future receive a more receptive hearing in the legislative chamber. The majority of proponents could and did take pride in the Copeland law. They understood that, viewed from the perspective of current political realities, they had gained a remarkably good statute.

The fourteen national women's organizations who had borne the brunt of the battle were delighted, and Ruth Lamb was quick to congratulate them on being the moving force in the final achievement.[74] Long-time reformers in the profession of pharmacy were also enthusiastic over the new law. *American Professional Pharmacist* called it a "mighty milestone" and applauded the profession for its role in bringing "order out of the chaos." Other pharmaceutical journals were equally pleased.[75] Even the *Journal of the American Medical Association* had a good word for the end result. The journal was something less than jubilant, but it did pay particular note to the advantages of provisions governing the introduction of new drugs into the public market. "It should not be necessary again," that periodical stated in restrained acceptance, "to propose the passage of an entirely new act."[76]

[74] "The New Food, Drug, and Cosmetic Bill and the Home Economist," *Jnl. of Home Economics* 30 (October 1938), 545-47. The fourteen national women's organizations so much responsible for ultimately gaining a new law were as follows: American Association of University Women, American Dietetic Association, American Home Economics Association, American Nurses' Association, Girls' Friendly Society, Homeopathic Medical Fraternity, Medical Women's National Association, Young Women's Christian Association, National Congress of Parents and Teachers, National Council of Jewish Women, National League of Women Voters, National Women's Trade Union League, District of Columbia Federation of Women's Clubs, National Women's Christian Temperance Union.

[75] "Strength," *Am. Professional Pharmacist* 4 (July 1938), 13; "The New Food and Drug Law," *AJPhE* 2 (July 1938), 396-97; "The New Food and Drugs Act," *AJP* 110 (June 1938), 222-25.

[76] "The New Federal Food and Drug Act at Last," *JAMA* 111 (July 23, 1938), 324-26.

Characteristic of the entire struggle, the passage of the act received limited coverage in the lay press. Five months after passage FDA was still getting letters from consumers and consumer organizations asking after the status of the food-drug bill.[77] Yet where the lay press did take notice the comment was generally favorable. Swann Harding's evaluation for the *Christian Century* was typical: "All things considered, this is a very good law." With obvious personal irritation Harding went on to add, "It is a far better law than putative consumer 'friends' who knifed the bill and attacked its sponsor at every opportunity, deserve."[78]

David Cavers, who had participated heavily in the drafting of the original 1933 measure, S. 1944, and who had served as a constant adviser to FDA on the successive revisions, was also content with the final product. It was not a perfect measure, he admitted, but he was "convinced that the new law represents a vast improvement over the old one, a gain well worth the five years of unremitting effort on the part of its Congressional champions."[79] On the occasion of the announcement that FDR had signed S. 5 into law, Secretary Wallace expressed much the same sentiments. There were some disappointing features, but on the whole the measure was a great step forward in the protection of the American public. The act will stand, Wallace concluded, "as a legislative monument" to Royal Copeland.[80]

This, perhaps, was the one real area in which there was room for regret. Copeland had fought hard and well. The proposed legislation was a matter very close to his heart. In the July issue of *American Perfumer* he explained the operation of the new law with obvious pride.[81] The article,

[77] Cavers, "Food, Drug, and Cosmetic Act," 3.

[78] "How Strong is the New Food Law," *Christian Century* 55 (June 29, 1938), 816.

[79] Cavers, 42.

[80] *New York Journal of Commerce*, June 29, 1938, and *New York Tribune*, June 28, 1938, FDA Scrapbooks, n.s., Vol. 1.

[81] Copeland, "New Food, Drug, and Cosmetic Law." *Am. Perfumer*, July 1938, Copeland Papers.

however, was published posthumously. Left among the New Yorker's personal papers was a small bit of poetry from an acquaintance in Massachusetts. It began, "I may grow old, perhaps, some day,/ But not in April—never!"[82] Copeland aged in April, indeed throughout the Congressional session. His doctors said he had driven far too hard in the past months. In the very last days of the Congress he had collapsed on the Senate floor. On the final night of the Congressional session he died—a week before Roosevelt signed the Senator's measure into law.[83]

Copeland did not live to see the formal culmination of his labors for food and drug reform. Yet, as *Printers' Ink* remarked on the occasion of the Senator's death, if in his last minutes he "was capable of coherent thought . . . he could have had much comfort out of the knowledge that he had done a complete, workmanlike and successful job up to the very end. His crowning achievement perhaps was the Food, Drug and Cosmetic Act. . . . And, for all its changes, it is now the Copeland law."[84] Here *Printers' Ink* was surely correct. The late Senator had doubtless foreseen with great satisfaction that the statute would carry his name.

[82] Undated poem from Emma Mosse to Copeland, Copeland Papers.
[83] *Washington Post*, June 18, 1938, FDA Scrapbooks, n.s., Vol. 1.
[84] "Cure Thyself," *PI* 183 (June 23, 1938), 80.

IX

ANATOMY OF REFORM

A perfect law has not been achieved. . . .
I am convinced that the new law represents a vast
improvement over the old one, a gain well worth
the five years of unremitting effort.

DAVID F. CAVERS
LAW AND CONTEMPORARY
PROBLEMS, 1939

THE New Deal as a complex of legislative and executive
actions dealing with major domestic problems in America
lasted only five years. Its force was gone by the end of 1938.
Traditionally those years have been divided into two periods, the First and Second New Deals, each with its respective characteristics. If this division be accurate, then
the 1938 Food, Drug, and Cosmetic Act is a curiosity. This
statute and the Fair Labor Standards Act represent the last
major domestic measures passed before foreign affairs began to absorb the chief energy of the Administration. In
time and in the nature of the legislation, it bridges both
New Deals. It does not, however, fit with ease into the standard conception of either period.

The history of the act began in 1933 with the Tugwell
bill. By standards of the business community, this bill was
an unpopular and harsh regulatory measure. Yet this was
a period in which the new Adminstration was presumably
seeking to combat the disasters of the Great Depression
through partnership with business in cooperative social
management. A partial key to this seeming paradox is in the
social philosophy of Rexford Tugwell, the driving force
within the Administration behind the proposed legislation.
Tugwell's conception of social management was compre-

hensive. In regard to business it rested on the gospel of efficiency. He envisioned the emergence of large productive units and the legalization of this combination movement. Yet he also demanded the establishment of extensive social controls to protect the general interest.[1] Planning meant much more than government's acting as an advisory agency. If there was to be partnership with business then government must be a senior partner, because business would not abdicate its traditional privileges voluntarily.[2] Tugwell was advocating the operation of an integrated group of enterprises primarily in the interest of its consumers rather than its owners.[3]

The Assistant Secretary of Agriculture was to the left of the majority of the architects of the First New Deal. His comprehensive ideas on planning found at best very limited expression. The dominant mood of the early Roosevelt years was best expressed in the National Industrial Recovery Act and the Agricultural Adjustment Act. These laws embodied the partnership concept and Tugwell was certainly not opposed to the approach. Yet the Assistant Secretary, and other Brain Trusters of his bent, did have mixed feelings particularly in regard to the NRA.[4] Tugwell was suspicious of businessmen and wary of a rationale which stressed scarcity profits. He was reluctant to entrust business with the operation of economic controls.[5] Government must be the dominant partner.

Moreover, if private enterprise was to be allowed the power of self-government under a relaxation of the antitrust laws then it owed something in return. There must be checks over matters such as excessive prices, unreasonable profits, and quality standards.[6] Consumer interests had to be protected. As a member of the Special Industrial Recovery Board, an interdepartmental agency charged with

[1] Bernard Sternsher, *Rexford Tugwell and the New Deal*, 109, 154.
[2] Tugwell, "Planning Must Replace Laissez-Faire," in Howard Zinn, ed., *New Deal Thought* (New York, 1966), 88.
[3] Ellis Hawley, *The New Deal and the Problem of Monopoly* (Princeton, N.J., 1966), 45.
[4] Sternsher, 339, 159.   [5] Hawley, 44.   [6] *Ibid.*, 74.

the responsibility of coordinating NRA with the rest of the recovery program, Tugwell again and again strongly objected to the establishment of price and production controls without adequate regulations in the interest of the consumer.[7] To him NRA administrator Hugh Johnson had not satisfactorily settled the question of relationships between government and business. Business had become the senior partner.[8]

It is in this sense that the Assistant Secretary's enthusiasm for a new food and drug act is best understood. NRA and even the AAA were lacking in the matter of consumer protection.[9] Tugwell hoped this situation would change. He had realized very early, however, that one avenue for regulatory action in the interest of the consumer was the Food and Drug Administration. He noted in his diary in February 1933 that FDA would be a first concern with him. He added with determination: "I'll do the best I can for the consumer regardless of politics; I won't compromise on this."[10]

In Tugwell's mind increased power for FDA would not seem inconsistent with the partnership approach of NRA. It would be the necessary regulatory side of social planning. Certainly the Assistant Secretary hoped for much more in terms of general governmental controls over economic operation, but a strong FDA could be viewed as at least one way of balancing out the partnership between government and business. Indeed, one might speculate that part of the violent reaction by the business community to the Tugwell bill was motivated by more than dislike for the particular bill itself. Tugwell and other New Dealers of like mind had fought to install strong consumer protection features into NRA. Perhaps business interests feared that extensive popular support for the food bill would increase the chance that rigorous provisions for protection of the consumer would become a part of the NRA operation.

It is of interest to speculate further on the reaction of

[7] *Ibid.*, 73.   [8] Sternsher, 161.   [9] *Ibid.*, 339.
[10] *Ibid.*, 225.

the affected business community in the light of Gabriel Kolko's 1963 study of the Progressive era. He argues that during that period at least, no legislation was passed which conflicted in a fundamental way with business supremacy over the control of wealth. Ultimately businessmen defined the limits of political intervention as well as specified its major form and thrust. Virtually every measure of importance to the business community passed in that period was not only endorsed by key representatives of the businesses involved but were first proposed by them.[11]

The Kolko thesis may go far to explain the history of what became the 1938 Copeland law. The original Tugwell bill was hardly proposed by the affected industries and assuredly did not have their sanction. Perhaps this very fact explains why the concerned economic interests were so shocked at the proposed legislation. Perhaps the bill was a threat not only to current commercial practices but also to their now traditional position as the major determinant in legislative change affecting the industries. Their reaction was to strike at the whole proposal in an effort to create a bill acceptable, if not totally desirable, to a business point of view. That effort achieved notable successes. In this sense the story of the 1938 food law was typical of much of the New Deal. Most of its major measures were significantly "watered down" before final enactment.

The affected industries may have had very solid reason to be shocked at the nature of the Tugwell bill. If Swann Harding's 1946 analysis of the Wiley law is correct, they may have been unwilling witnesses to a potential revolution. Harding asserts that the 1906 act was not generally recognized by the business community as primarily consumer legislation. Rather it was understood as a measure to ensure fair trade practices and to eliminate unfair competition. Even Harvey Wiley, for all his verbal ferocity, tried to form a regulatory policy which would disturb business as little as possible. Only gradually did FDA develop

[11] Gabriel Kolko, *The Triumph of Conservatism* (Chicago, 1967).

into an agency whose main concern was consumer protection. This change caused no small anxiety in the affected industries. Thus they could see the effort for increased power in the agency as strengthening and perpetuating a dangerous innovation. FDA was, as one journal warned its readers in 1934, "a group of people prejudiced in the interest of the public."[12]

The 1938 Food, Drug, and Cosmetic Act was passed into law after the presumed political shift to the left, characteristic of the Second New Deal. On one level the measure fits well. A strong regulatory agency with powers to provide greater security and justice for American consumers was complementary to the new Brandeisian posture toward the business community. Even Rexford Tugwell, who was less than enthusiastic about the shift in policy, believed this to be the case.[13] He urged that a strengthened FDA was "strictly in the line of the new philosophy—to regulate industry, but not to require of it planning or performance. If manufacturers were to be required to treat each other fairly, they ought also to be asked to treat their consumers fairly."[14] Thus the proposed food-drug law, without change in nature, easily bridges the two New Deals.

Yet on closer scrutiny the history of the Copeland law does not rest comfortably in the stereotype of the Second New Deal. If what marks that period was a more militant approach toward the business community, then in a real sense the Copeland bill moved right as the New Deal moved left. Helen Sorenson argues in her 1941 volume, *The Consumer Movement,* that "the particular form given to the new law was to a considerable extent a test of the forces lined up to oppose the reform."[15] The assertion might

[12] T. Swann Harding, "The Battle for a Better Food and Drug Act," *AJP* 118 (October 1946), 341.

[13] Tugwell denied that the New Deal made a legitimate turn to the left. He saw it as a conservative turn. See Paul Conkin, *The New Deal* (New York, 1967), 57.

[14] Sternsher, 122.

[15] Maxwell Cleland, "The Guinea-Pig Muckrakers: Proponents of a Consumer-Oriented Society" (Master's thesis, Emory University, 1968), 77.

well bring a knowing smile from Professor Kolko. Miss Sorenson's view is not without important substance. The process of compromise with the affected trades began early. Each industry received significant concessions. The greatest trade victories, in FDA's view, came in 1935 and after. Notable among these were more lenient regulations on label information, reduction in the power of multiple seizures, investment of advertising controls in the Federal Trade Commission, and the addition of elaborate court review provisions.[16]

As the Copeland bill passed into law it did represent something akin to the business consensus Kolko believes typical of legislation in the Progressive period. Indeed it seems quite fair to suggest that without the increasing body of trade interests brought into the consensus after 1935 no federal food and drug measure would have passed into law before the opening of World War II. The full story of the growing support by the business community for passage of the Copeland bill is a complicated one, and has been discussed in detail in the preceding chapters. One primary point, however, should be recalled here. The support by many segments of the affected trades was forced as a defensive reaction.

The movement for a new law was not primarily rooted in interest desires of industry as had been the case during the Progressive era. As James Harvey Young has explained, "competition was less a force [in precipitating the new control bill] than during the period prior to 1906. At that time the basic bills had originated with farm and industry groups which had no national protection against debasers. Now [in the New Deal] the draft bill came from the regu-

---

[16] On the matter of advertising controls the trade was of course divided and in great part the shift was a result of efforts by FTC. Still, this was a change greatly sought by some segments of the affected trades, notably the National Association of Wholesale Druggists, the Institute of Medicine Manufacturing and individual members of the Proprietary Association. This writer believes that, given the state and history of FTC versus FDA in the decade of the thirties, the investment of advertising controls in the former did represent a loss in consumer protection.

lating agency, concerned more with consumer protection than with business competition." Young concludes that the major competitive element involved in the struggle was between the Federal Trade Commission and the Food and Drug Administration.[17]

In attempting to place the Copeland law within the context of the New Deal it is vital to understand that the movement for a new law, as well as the text of the proposed measure, did originate in the Food and Drug Administration. The original text was not the product of New Deal philosophy per se, nor the thought of Franklin Roosevelt or even of Rexford Tugwell. The natural tendency of contemporaries to view it as a part of the existing political context may have been harmful. FDA's Ruth Lamb felt so. "It is unfortunate in some ways," she wrote in 1936, "that this much-needed revision of the Pure Food Law has been in the legislative hopper along with other New Deal measures so controversial in the same respect." To Miss Lamb the proposed new law carried no political philosophy; it was designed simply as "a practical method of coping with ever-changing technical problems, a method predicated on nearly thirty years of administering an inadequate law."[18]

If the chances of a new food and drug law were possibly hurt by association with New Deal theory, they were assuredly not helped by Roosevelt's general perspective on legislative priorities. Relief and recovery always took precedent over reform in the New Deal. Food legislation was of course a matter of reform having only indirect reference, if any, to the other R's. Reminiscing on the depression years, Rexford Tugwell recalled that this very fact posed a considerable problem when he urged the President to use influence on behalf of a food-drug law. Roosevelt reminded Tugwell that the Assistant Secretary himself had been a prime voice insisting that a careful distinction must be made between reform and recovery. Reform matters

[17] Young, "Social History of American Drug Legislation," in Paul Talalay, ed., *Drugs in Our Society*, 225.
[18] Lamb, *American Chamber of Horrors*, 291-92.

should be considered only after recovery was under way. Thinking further on the subject Tugwell said with a smile, "I answered that, well maybe because I didn't like your reforms."[19]

There were related political factors which also affected the low priority given food-drug revision. Once the proposed legislation was introduced, as Tugwell put it, the President came in for considerable "political flack" and for no small number of warnings from Congressmen that the "disturbance" created over this matter was harmful to the early passage of other programs. The Assistant Secretary admitted that even he had some second thoughts when the bill became such a cause with Congressional conservatives. There were too many other things that needed doing.[20]

The President was a political realist. If food and drug legislation jeopardized the recovery operation, then, regardless of his enthusiasm for such legislation, FDR would give it little support. He frankly admitted such practical considerations in regard to action carrying greater import than food law revision. In 1935 Walter White, Secretary of the NAACP, urged the President to make civil rights legislation a part of the New Deal. Roosevelt replied quite simply that he could not "take the risk" involved. If he came out for such legislation, FDR stated, "they [Southern Congressmen] would block every bill I ask Congress to pass to keep America from collapsing. . . ."[21]

The matter of general priorities was not the only factor involved in the Presidential relationship to food-drug law revision. Quite clearly Roosevelt had little personal enthusiasm for that reform. His initial involvement and limited concern throughout the course of the legislative battle seemed to flow more directly from the influence of his wife Eleanor and Rexford Tugwell than from any belief in the

[19] Tugwell, Rexford G., An Interview with Charles O. Jackson, June 7, 1968. Transcript in Oral History Collection, National Library of Medicine, Bethesda, Maryland.
[20] *Ibid.*
[21] Monroe Berger, *Equality by Statute* (New York, 1952), 15.

urgent necessity for greater regulation over the food, drug, and cosmetic market. Because the President was apathetic in this matter he allowed the course of Copeland-sponsored bills to be affected by his own bad relations with the New York Senator. Publicly his support for revision was always limited to calls for passage of "a bill" rather than "the bill." This fact encouraged industry opponents to press for weakened provisions or even weaker bills than Copeland and FDA believed adequate. If FDR did not actively seek to use the Senator's keen interest in revision as a basis to punish him for his political sins, the Presidential attitude did encourage other party stalwarts to do so.

Because the Chief Executive was apathetic to food-drug reform, and displayed this attitude publicly, members of Congress were encouraged further to believe that delay in passage of a bill as well as the seeking of concessions in the interest of industries "at home" was acceptable activity. Because the revision effort was a minor concern with the President he chose to deny the responsibility of leadership. He provided few guidelines to the Congress as to what he considered most adequate in the protection of consumer interest. The advertising control issue was a prime example. Roosevelt should have indicated some preference between the contending executive agencies. His failure to make a choice, or to make a choice clear, helped to create a major controversy where one might have been avoided.

Part of the Chief Executive's apathy might have been changed if the respective Copeland bills had attracted strong and unified consumer support. Such was not the case. The majority of the public was seldom, if ever, drawn into the reform camp. It is pertinent to note, however, that opponents of legislative revision suffered from the same malady. They too were never able to create prolonged opposition to food-law reform among the public. Each side had its moments. The opposition was most successful during 1933-1934 when dealing with the Tugwell bill and its first major revision. Sections of the public were receptive to an appeal based on violation of American traditions. Proponnents of reform gained their highest degree of support during 1937-1938 in the wake of the Elixir Sulfanilamide tragedy. The specter of death made reform for the time a personal matter. In the interval both sides were hard put to attract popular attention.

This apathy, while especially unfortunate for the food and drug reform drive, is not difficult to understand. One factor was the Great Depression. Widespread economic troubles—very personal matters of day-to-day existence—set natural limits to the activities and interests of much of the public. People might look with keen concern to the activities of the government but, for most, concern was with two *R*'s, not three. With the public, even more than with the President, reform was bound to take a back seat to relief and recovery. A second consideration is the nature of food and drug reform itself. Clearly it is—except during episodes of crisis—"dull stuff" for the vast majority of the American people.

Basic issues in such reform are often technical or confusing and tend to sound totally abstract. Matters like minute word changes in phrasing of multiple seizure power, appeal procedure from regulations by the Secretary of Agriculture, or even dispute over which government agency should control advertising exasperate, rather than inspire ardor in, most people. Such issues, however crucial, lack the color of FDR's court-packing fight or even the issue of Tennessee Valley development. Only when the New Deal food and drug fight could be drawn in dramatic terms—socialism versus democracy in 1933-1934 or life versus death in 1937-1938—would the public respond. Thus the major part of the long struggle over a new food law was carried out within a restricted body of contestants.

The possibility of much stronger consumer protection in the Depression decade was threatened by problems other than public apathy. As in any responsible democratic government, the New Deal sought to allow every interest a voice in the direction of legislative movement. Consumers were generally disadvantaged, however, in this "interest-

group" democracy. At least in the Roosevelt years the amount of voice each interest had was largely dependent on its organizational power, and this the consumer lacked.[22] Moreover, even those organized groups concerned with greater consumer protection in foods and drugs lacked unity. Ironically, their respective conceptions of self-interest kept them partially separated. To that degree early passage of a new law was retarded.

One potential part of the reform coalition was the American Medical Association, which had played a major role in the passage of the 1906 Wiley law. As the spokesman of the nation's medical profession and a powerful political pressure group the organization could have provided invaluable aid in the battle for what became the 1938 act. That assistance came only in a minimal way. The AMA supported revision in the abstract but never gave more than qualified endorsement to any Congressional bill. It was hypercritical of all. The failure to take an active part in the reform campaign can only be viewed as ambivalence in self-interest. Restriction on medical quackery and pseudomedicine was a concern of the organization. Yet, it appears that controversy of recent years over such matters as public health insurance and the cost of medical care had promoted a defensive attitude on the part of the AMA. That body became fearfully reluctant to join any legislative movement which might open the door to federal intrusion in the field of medicine. So the organization chose to support revision in principle but to deny concrete association with any specific measure—a posture of little help in passing the Copeland bills.[23]

A second potential membership in the food-drug reform movement was the professional consumer organizations such as Consumers' Research and Consumers Union. These groups certainly felt that statutory revision was a matter

[22] Cleland, "Guinea-Pig Muckrakers," 46.

[23] Some contemporaries have suggested that the posture of the AMA may have been partly a result of the fear of losing drug advertising in their publications. This does not seem to have been a significant factor in the Depression decade.

of vital interest. They worked hard to press this recognition on the public in a variety of ways. Yet by the nature of these groups—because consumer protection was itself a business—they tended to function as the reform counterpart of the most militant resistance to revision in the affected industries. Their conception of adequate consumer safeguards which might be passed into law was frankly utopian. The consumer-oriented society with its ideal goals of perfect manufacturing standards, absolute truthfulness in advertising, and flawless extensive controls over the food-drug market was not a real possibility even in the revolutionary ferment of the thirties.[24]

One recent appraisal of the professional consumer movement in that decade offers the interesting judgment that the Depression made governmental concern for quality standards in consumers goods seem a luxury. The issue of material want and recovery had to dominate the politics of the New Deal. The very importance of increased consumption in the process of recovery dictated that new restrictions on industry, especially legal demands for more truth in advertising, would be applied only with great hesitation. The important thing was to get the consumer to consume. The professional organizations did not comprehend this fact. Rather, men like F. J. Schlink regarded the hesitation and the concessions to industry as simply another dramatic illustration of Congressional "sellout" to the business community.[25]

Whether the militant view of adequate consumer protection held by these organizations was the right of consumers or not is here an academic point. In fairness it might be noted that their rigid posture may have served as a significant counterbalance to the hard right of the food-drug trades. Thereby they may have helped to halt more serious compromise in the respective Copeland bills. Yet, more important, the interest-conceptions of these groups put them at odds with the ideas of Royal Copeland and the views

[24] Cleland, 45.
[25] *Ibid.*, 44-45.

of FDA. Thus these organizations declined to support the main line of the revision effort—the effort that did have a chance to succeed—and encouraged their followers to do the same. Early enactment of a new food law was thus made more difficult.

The most effective support for the FDA version of reform came from the national women's organizations. This support was supplemented by a limited number of periodicals, such as *New Republic, Nation, Christian Century,* and *Saturday Review,* as well as significant segments of professional pharmacy. Without this backing it seems certain that no new law would have passed. The women in particular were quite convinced that a new law was in their personal interest. Their view of adequate consumer protection was less utopian than the professional consumer organizations and they took a major role in the struggle to pass the successive Copeland measures. They maintained a "realistic" understanding of the legislative process. As a result their position underwent the usual processes of goal modification and attrition which generally characterize legislative reform battles. Yet, their central if unspoken interest was to keep concessions to a minimum and to get some law on the books which would eliminate the worst abuses in the marketplace. Their constant pressure on the opposition was the effective force which finally enabled a new law to pass.

The opposition to revision of the 1906 statute was initially a very formidable thing. As David Cavers pointed out in 1939, not only were the affected industries as a whole huge in size but they were also decentralized. There were the giants, but typically concerns were moderate and small in size and so well distributed over the country that every Congressman had in his constituency at least some affected business.[26] These industries were opposed to revision, particularly a totally new law, if for no other reason than the uncertain consequences of change. It ap-

[26] Cavers, "The Food, Drug, and Cosmetic Act of 1938," *Law and Contemporary Problems* 6 (Winter 1939), 4.

pears likely, given the limited degree of consumer interest in revision, that a solidly united food, drug, cosmetic, and advertising business community could have blocked passage of any significant change in the law. The stumbling block to continued united and militant opposition was the matter of self-interest.

From the start of the legislative battle the food industry posed a problem for those who sought to halt the coming of a new law. Since that industry as a whole was the least affected of all the trades, even by the Tugwell bill, its leadership was far less interested in militant resistance. What they really wanted was a separate food bill. They based their argument in this regard on the assertion that their interests were different from the other trades. As for the Tugwell measure, food representatives were opposed to full formula disclosure, multiple grading of food products, and the wide powers to make regulations given to the Secretary of Agriculture.

The industry did not get a separate law. In a short time, however, full formula provisions were dropped, along with multiple grade provisions. Advisory boards to make less discretionary the decisions of the Secretary were added. There were other lesser features of the successive bills which the food men would have liked altered and which were not changed. Yet these additional desires had to be measured against both the major concessions received and an uncertain future in which more objectionable features might be added to reform proposals. By the end of 1934 food processors had made their decision and were dropping out of the opposition coalition.

A similar process of interest evaluation took place in the advertising industry, which certainly was more heavily affected than the food industry by the Tugwell measure. This fact, plus the huge total in advertising dollars expended by the food, drug, and cosmetic business community, doubtless helped promote the "bad press" which revision efforts received. Advertising spokesmen were pleased, however, when subsequent bills omitted multiple grading provisions which they had felt would cut into advertising revenue. They were also pleased when publishers who unknowingly accepted false advertising were specifically exempted from prosecution under the advertising sections of the subsequent Copeland bills.

They were not pleased that by 1935 many advertisers were balking at signing new contracts until the matter of a new law was settled. They were less than happy over the continuing criticism of advertising which came as a corollary to the continued argument over a new law. They too were fearful about the future. Sixty bills touching advertising had been introduced in Congress by September 1935. Generally the industry was not pleased that advertising control might go to the FTC rather than FDA. At length the balance of self-interest turned in a different direction. By the end of 1935 that industry too began a withdrawal from the opposition councils.

The most prolonged and adamant resistance to legal revision came from the drug-cosmetic interests, because that area of business was most affected by the Tugwell bill and its successors. There were significant divisions, however, in the ranks of this group. The primary division was between the more professionally minded organizations, such as the American Association of Colleges of Pharmacy, versus the more commercially oriented groups, such as the National Wholesale Druggists Association. In general, the professional groups initially stood united with other drug interests in opposing the Tugwell bill. This soon changed. The American Association of Colleges of Pharmacy, the American Pharmaceutical Association, and organizations of similar type had a major interest in the scientific aspects of drugs as well as in public health. They were also interested in establishing pharmacy as a recognized profession. They were less concerned with the commercial side of the business and they were relatively immune to the pressures of the drug manufacturers.

These groups were in fact predisposed to support stronger controls over drugs. Their initial opposition sprang more from a natural resistance to sudden change and a desire for unity among drug interests rather than any serious objection to tighter restrictions on the trade as such. Moreover, the basic legal changes would affect manufacturers more than professional pharmacy. The changes also had greater consequence for patent medicines than for prescription drugs. There was no strong reason why these more professional groups should seek very hard to advance the cause of proprietary medicines. The growth of that industry did not really enhance the status of pharmacy. After all, proprietary mixtures might be sold over many counters other than that of the pharmacist. Finally, in 1934 some of these professional groups came to feel that they had been hurt by allowing themselves to be associated with the antics of the militant drug manufacturing bodies. By 1935 a number of the more scientific organizations were calling for passage of the Copeland bill as it stood at that moment. Some, like the American Association of Colleges of Pharmacy, carried on an active campaign for a new law.

A second division among drug interests was between prescription and proprietary drug producers. This interest division should not be overestimated for, as Swann Harding has pointed out, prescription companies were also involved in the production of proprietary products.[27] With this caution in mind, however, it is valid to note that the views of the two types of producers were not always the same. The very nature of the prescription business meant that producers must be more scrupulous about appearances and show more interest in medicine control and research than the proprietary men. The prescription industry must advocate better legislation if only for the sake of appearances.[28]

Appearances aside, logic early suggested to many prescription producers that the proposed restrictions on proprietary products and their sales might well work to their advantage. What did not work to their advantage was the bad publicity which the controversy over revision brought

[27] Harding, "Battle for a Better Food and Drug Act," 343.
[28] *Ibid.*

upon the whole drug industry. Most of the reform exposé material was aimed at the proprietary end of the business but laymen in reading such material would not make a distinction. Medicine was medicine.

This is not to suggest that prescription drug producers were ever very enthusiastic about a new law. They remained much concerned about a number of provisions of the proposed reform such as the variation clause and later the power given to FDA to control the safety of "new drugs." In the end they, like their proprietary colleagues, reluctantly accepted passage as less dangerous than seeing the struggle continue. Apparently because there were interest differences between themselves and the proprietary business, the prescription industry, however, never took the overt active role in resisting revision that they might have taken. They chose to let their proprietary brothers carry the brunt of the battle.

Proprietary drug and cosmetic manufacturers were the group most affected by the drive for reform, particularly with regard to advertising controls—to them a crucial matter. In 1933 they undoubtedly hoped to block completely any new bill. Only by the end of 1934, with continued pressure from groups like the women's organizations and with the gradual capitulation in other segments of the affected industries, did some proprietary manufacturers come to believe some new law was inevitable. As producers recognized this fact their tactics changed from destruction of the reform drive to delay and to making that new law as weak as possible. Some segments of the industry, of course, especially manufacturers whose products had the largest sales, made this modification with more celerity than others. In the course of the struggle there were significant victories for the proprietary legions. Prominent examples were the modification of multiple seizure powers and (at least as viewed by a portion of the industry) the shift of advertising control power to FTC. The drug men might have fought on indefinitely if the real and potential cost of battle had not risen with the years. There was the fact that by 1936 they stood almost alone in their battle. There was the anxiety about what each new revision of the bill would include and the certainty that new bills would come. There was the ever growing body of criticism of the medicine market spread over the land by the "guinea pig" muckrakers.

By 1936-1937 there was the very real threat of state and local legislation on drugs, divergent in their provisions, which was viewed as more dangerous to the well-being of the industry than a federal statute. Then came the Elixir Sulfanilamide disaster, which for the first time really brought a heated public demand for legislation. Even temporary public wrath was bad for business. It reduced the number of legislators willing to stand adamantly for the interests of the industry. In short, by 1938 the cost of battle was too high. Industrial self-interest now dictated that a law should be passed. Attrition had taken its toll. So in June 1938 there was a new food and drug law.

Interest conflict is basic to the American legislative process. The law-making procedure does respond to divergent pressures and because legislative stalemate is not a satisfactory form of conflict adjustment, end results tend toward compromise measures. In this regard the history of the Copeland law was the typical Congressional story. The final product was not a perfect law. Significant losses in consumer protection were involved in the many concessions made to affected industrial interests during the long struggle for a new statute. Dependent on perspective, evaluation of the statute can take at least two very different paths. In the sense that more was not done for the consumer when so much more was needed, the passage of the law could be viewed as socially abortive. Certainly groups like Consumers Union felt so. As late as 1957 Rexford Tugwell would still insist that the law as passed was "a discredit to everyone concerned in it."[29]

Yet perspective is the all-important word. Tugwell's perspective was that of a defeated social planner. His extensive

[29] Tugwell, *The Democratic Roosevelt*, 464.

philosophy of social management was an alternative program never implemented by the New Deal. In the last analysis Tugwell's dissatisfaction had little to do with what happened to a single bill which originally bore his name. Rather the compromises embodied in the Copeland law represented one of many rejections of his basic planning concepts. Consumers Union's perspective was as an agency demanding a change from a business-oriented society to a consumer-oriented society. Here too the compromises in the 1938 statute represented the rejection of a philosophical position. The historian of similar leanings will also find much to criticize about the Copeland law, for the matter of perspective applies equally to the scholar. From this standpoint the failures of the statute are related to the political conservatism of the New Deal which encouraged only moderate socioeconomic change.

Yet, a second view of the 1938 act is also possible. In the month of June the food measure and the Fair Labor Standards Act passed into law. As previously noted they were the last major domestic measures passed by the administration. "The New Deal, which began in a burst of energy," to quote historian Paul Conkin, "simply petered out in 1938 and 1939."[30] Roosevelt's support for domestic reform waned as his concern for foreign affairs increased. The food and drug fight spanned the entire New Deal period and achieved concrete results only at the very end of it. Unquestionably the final legislative product was tainted by significant concessions to the affected business community. The alternative to passage in 1938, however, would have been an indefinite delay for any revision of the long-outdated Wiley law. Perhaps the needed changes would not have come until after World War II or until other and more ominous disasters on the order of the Elixir Sulfanilamide tragedy demanded action. Considering the extent of the chemotherapeutic revolution then taking place such disasters might have come on a large scale.

Realists of the day like David Cavers were very clear

[30] Conkin, 101.

that the new law was not perfect. They simply argued that it represented a vast improvement over the old statute.[31] The affected industries had not been ousted from their position as a primary formulator of legislative policy touching their interests. The American business community, indeed, retained that role in the New Deal as a whole, whatever the rhetoric of the administration. Yet, characteristic of American reform, the worst conditions needing control were covered by the provisions of the new food law. Moreover, in retrospect it would seem that both industry and reformers greatly underestimated just how effective the Copeland measure would turn out to be.

The basic worth of the statute has been greatly enhanced in the years since 1938 by forceful and imaginative administration by FDA. Judicial interpretation has also tended to strengthen and broaden the law. A case in point was two decisions by the Supreme Court in 1948. In Kordel v. United States and United States v. Urbuteit the high tribunal held that product literature which met the product at the location of sale was labeling. It was thus subject to the labeling provisions of the law even though such material did not physically accompany the product.[32] The statute also laid significant foundations for the future. Thus the new-drugs provision offered precedent for later legislative action based on the concept of preventive protection—putting the burden of safety, and eventually of efficacy in some areas, upon the manufacturer prior to marketing.

Time has disclosed weaknesses in the 1938 statute but additional change has proved possible. In this case Congress has not allowed the act to become so outmoded as to demand a new total-revision effort. Significant amendments have been made since 1938.[33] The structure of legislation

[31] Cavers, "Food, Drug, and Cosmetic Act," 42.

[32] Walter F. Janssen, "FDA since 1938," in The Government and the Consumer, 29.

[33] These changes include: the Miller Adulteration Amendment (1948), Humphrey-Durham Drug Prescription Act (1951), Factory Inspection Amendment (1953), the Pesticide Chemical Act (1954), the Food Additive Amendment (1958), Color Additive Amendment

of laws designed to protect the American consumer is still far from perfect but the present-day consumer is at least better protected than ever before. The Copeland statute remains a very basic element in this structure.[34] As such it should be regarded as a major accomplishment of the New Deal years.

(1960), Kefauver-Harris Drug Amendments (1962), Drug Abuse Control Amendment (1965), Fair Packaging and Labeling Act (1966).

[34] For a brief statement of trends in government control over the food, drug and cosmetic market since the passage of the Copeland law see Janssen, "FDA since 1938," 21-37.

BIBLIOGRAPHICAL ESSAY

THE following bibliographical comments are not an all-inclusive statement of materials used in the preceding study. Rather the remarks here made are designed to indicate major sources, to provide a brief evaluation of these materials, and to call special attention to those sources of particular assistance.

SECONDARY MATERIAL

Secondary sources bearing directly on the struggle to pass a new statute or on the 1938 law itself were limited. Even monographic studies on the New Deal provided only the most cursory and abbreviated coverage of this legislation. Brief but valuable commentary of Rexford Tugwell's part in the drive for a new law was found, however, in Bernard Sternsher's *Rexford Tugwell and the New Deal* (New Brunswick, N.J., 1964). Of more limited value in its statement regarding drug revision efforts was Arthur M. Schlesinger, Jr.'s *The Coming of the New Deal* (Boston, 1959).

Several more specialized works proved useful on specific aspects of the story. Otis Pease's *The Responsibilities of American Advertising* (New Haven, 1958) provided important insight into the position of the advertising industry relative to the drug revision effort as well as on the entire matter of federal regulation over advertising. James Burrow's *AMA: Voice of American Medicine* (Baltimore, 1963) provided valuable information relative to the position of organized medicine on drug reform. Extremely significant background information on the struggle for the 1906 food law, on Harvey W. Wiley and the early days of FDA was found in Oscar E. Anderson's *The Health of a Nation* (Chicago, 1958) and James Harvey Young's *The Toadstool Millionaires* (Princeton, 1961). These scholars, along with Wallace F. Janssen of FDA and Professor A.

Hunter Dupree, have each contributed useful articles bearing on federal control over the food and drug market in a pamphlet titled *The Government and the Consumer: Evolution of Food and Drug Laws* (Chicago, 1962); the Anderson, Young, and Janssen articles also appear, in slightly revised form in *Journal of Public Law,* XIII (1964), 189-221. This writer has benefited also by reading Professor Young's article, "The Elixir Sulfanilamide Disaster," *Emory University Quarterly,* XIV (December 1958), and portions from his recently published work on American health quackery in the twentieth century, *The Medical Messiahs* (Princeton, 1967). Valuable insight into the views of the professional consumer organizations during the New Deal was gained by reading Joseph Maxwell Cleland's "The Guinea-Pig Muckrakers: Proponents of a Consumer Oriented Society," M.A. thesis, Emory University, Atlanta, 1968.

PRINTED AND MANUSCRIPT SOURCES

While secondary material on food-drug reform in the New Deal was limited in nature, pertinent primary material was quite abundant. Basic to the examination of revision efforts was the extensive incoming and outgoing correspondence of the Food and Drug Administration for the years 1933 through 1938. The bulk of this material was located in the Economic and Social Section of the National Archives, Washington, D.C., under the reference title Accession No. 2468, Records Group 88. The most pertinent part of this very large collection on specific developments in the history of the respective Copeland bills was Decimal File .062, Correspondence on Legislation. This material for the year 1938 was deposited in the Federal Records Center, Alexandria, Virginia, under Accession No. 52-A89.

Also vital to the study and a part of Records Group 88 in the National Archives were the Records of the Commissioners of FDA under the Accession No. 52-A86. This collection was made up of some thirty odd boxes of letters, memos, and miscellaneous items from the files of the commissioners. Especially noteworthy were a number of scrap-books prepared in FDA during the 1930s. These books allowed this writer an otherwise practically unattainable view of the nationwide trade and lay press reaction to the course of the reform battle. No less valuable in understanding the political intricacies of the 1938 law's history were the Franklin D. Roosevelt Papers, more particularly Official File 375, relating to the Food and Drug Administration, in the Roosevelt Library, Hyde Park, New York. Any realistic appraisal of the President's role in the revision effort would be impossible without consulting this correspondence.

Of less significance, but nonetheless necessary to the overall picture of the movement of the Copeland bills through the halls of Congress, were the Petitions and Memorials Files of the Senate and House Commerce Committees which handled the food and drug bills. The Senate file was particularly fruitful in that it contained a body of correspondence from Royal S. Copeland to colleagues and other interested parties on the progress of the proposed legislation. These files are located in the Legislative Branch of the National Archives, Washington, D.C. The Senate material is in Records Group 46, the House file in Records Group 233. The Records of the Office of the Secretary of Agriculture (Rexford Tugwell) also proved of limited value. Particularly noteworthy in this collection was an analysis of the state of the food-drug market in 1933 and the shortcomings of the 1906 statute made by the Consumers' Advisory Board of the National Recovery Administration. This material in the Agricultural Section of the National Archives is in Records Group 16, Accession No. 1074.

Treatment of the Elixir Sulfanilamide disaster was principally based on three sources. The basic facts were set out in Senate Document No. 124, 75th Congress, 2nd Session, Report of the Secretary of Agriculture on Deaths Due to Elixir Sulfanilamide-Massengill. The complete story, however, required careful scrutiny of two large FDA files upon which the official report was based. The files were Sulfanilamide, Chronological File on Reports of Deaths, AF 1-258,

Sub-file 510-.2055, and a two-volume file titled Sulfanila-mide–S. E. Massengill Company, AF 1-258. Both are a part of Records Group 88 at the Federal Records Center, Alexandria, Virginia.

The *Congressional Record* was, of course, a basic research source in following the legislative history of the successive Copeland bills from 1933 on. A handy reference to points of Congressional debate was Charles W. Dunn's *Federal Food, Drug, and Cosmetic Act: A Statement of Its Legislative Record* (New York, 1938). Dunn reprints all pertinent portions of House and Senate debate along with specific references to the *Congressional Record*. Five sets of Congressional hearings proved invaluable to this study. One of these hearings took place in 1930 and thus precedes the actual battle for the 1938 law. It was, however, quite beneficial in understanding the limitations under which FDA labored in the movement for food-drug reform in 1933. The citation for this material is *Administration of the Food and Drugs Act, Hearings before the Committee on Agriculture and Forestry,* U.S. Senate, 71st Congress, 2nd Session, Washington, D.C., 1930.

The four additional hearings took place at various points in the first two years of the revision effort. Their importance lies in clarifying the specific desires as well as grievances of both proponents and opponents of reform. They were further useful in providing insight into trade strategy to block or water down the respective bills. These hearings were as follows: *Foods, Drugs and Cosmetics, Hearings before a Subcommittee of the Committee on Commerce, U.S. Senate . . . on S. 1944,* 73rd Congress, 2nd Session, Washington, D.C., 1933; *Foods, Drugs and Cosmetics, Hearings before the Committee on Commerce, U.S. Senate . . . on S. 2800,* 73rd Congress, 2nd Session, Washington, 1934; *Foods, Drugs, and Cosmetics, Hearings before a Subcommittee of the Committee on Commerce, U.S. Senate . . . on S. 5,* 74th Congress, 1st Session, Washington, 1935; *Foods, Drugs, and Cosmetics, Hearings before a Subcommittee of the Committee on Interstate and Foreign Com-*

merce, U.S. House of Representatives . . . on H.R. 6906,
H.R. 8805, H.R. 8941, and *S. 5,* 74th Congress, 1st Session,
Washington, 1935.

PRIVATE PAPERS

In addition to the official sources indicated above, scrutiny of three sets of private papers proved profitable in varying degrees. Most beneficial were the Papers of Josiah W. Bailey in the Duke University Library, Durham, North Carolina. This was an extremely large collection loosely cataloged in the file system used by Bailey himself. The most useful portion was the materials filed under "Department of Agriculture." Bailey was a major opponent of revision and an astute spokesman for proprietary medicine interests. His papers had value in explaining the Congressional strategy of the opposition as well as providing a clear look at the type of pressure faced by Congressmen whose constituencies included heavy trade interests.

Of less value but still pertinent were the Papers of T. Swann Harding located under the reference, LC III-23-P,3 in the Manuscript Division, Library of Congress, Washington, D.C. Harding was a one-time employee of the Agriculture Department and a devoted advocate of stronger food-drug controls. He was also a professional writer on scientific topics and much of the collection was not pertinent to the specific issue of revision in the 1930s. Scattered throughout the material, however, were pieces of correspondence and other items which were quite useful in piecing together the involved history of the 1938 law.

The Papers of Royal S. Copeland, in the Michigan Historial Collections, University of Michigan, Ann Arbor, had some value. At the time this collection was consulted the correspondence portion of the papers actually related more to the Senator's private life than to his public life, but the material did provide some pertinent information on the personality and early background of the man. The Michigan collection further included some thirty-five very large scrapbooks mainly of press material on Copeland's public career from 1889 through 1937. Perhaps because press coverage of the food-drug reform effort was so limited the pertinent volumes contain only minimal material relating to the food-drug law. The volumes were useful, however, in understanding Copeland's posture within New Deal politics as well as his political differences with Franklin D. Roosevelt. Since the present volume has gone to press Michigan has acquired a very large collection of Copeland's correspondence relating to his legislative career.

PERIODICAL AND OTHER PUBLISHED MATTER

It is impossible to overestimate the importance of periodical and other published matter to any understanding of the whole issue of revision in the 1930s. Only here can the researcher fully follow that dimension of the revision battle carried on by protagonists outside the legislative halls—the calls to arms, the fears, the rumors, and the judgment of events by concerned groups within the public. The periodical coverage, even in the trade press, was brief and no specific references were so valuable as to warrant special mention in these bibliographic comments. Specific references are, of course, abundant throughout the text of the study. In general terms, all of this material falls naturally into four categories.

One category was the large body of "guinea pig" muckraking literature. These books, of which Stuart Chase and F. J. Schlink's *Your Money's Worth* (New York, 1928) was the prototype, provided a continuing critique, from 1933 on, of the food, drug, cosmetic, and advertising industries from the point of view of militant consumer interests. Beginning with *100,000,000 Guinea Pigs* (New York, 1933) by Arthur Kallet and F. J. Schlink, the number of such publications grew yearly throughout the period of the struggle for the new law. The large number of these volumes prohibits a full listing here but several of the more pertinent and typical of these works do require mention.

Two "shockers" in 1934 were James Rorty's *Our Master's Voice* (New York, 1934) and M. C. Phillips' *Skin Deep, The Truth about Beauty Aids* (New York, 1934).

The most explosive volumes of 1935 were *Counterfeit— Not Your Money but What It Buys* (New York, 1935) by Arthur Kallet, and F. J. Schlink's *Eat, Drink, and Be Wary* (New York, 1935). The following year brought Joseph Matthews' *Guinea Pigs No More* (New York, 1936) and Rachel Palmer and Sarah Greenberg's volume, *Facts and Frauds in Woman's Hygiene* (New York, 1936). The year 1937 saw such works as Peter Morell's *Poisons, Potions, and Profits* (New York, 1937), Ruth Brindze's *Not To Be Broadcast* (New York, 1937), and Rachel Palmer's *40,000,000 Guinea Pig Children* (New York, 1937). Typical of the next year was George Seldes' *Lords of the Press* (New York, 1938). The most rewarding location for investigation of the "guinea pig" literature was the Library of Congress, Washington, D.C.

One additional volume not properly a part of the guinea pig group but closely related to them requires special mention. That volume is Ruth deForest Lamb's *American Chamber of Horrors* (New York, 1936). Miss Lamb was FDA Information Officer and her book had a twofold significance. The book itself was a highly popular and sensational addition to the guinea pig critique of the concerned industries in 1936. Secondly, it was a valuable piece of commentary from the point of view of the FDA as to exactly why and how the Copeland bill was still locked up in Congressional halls after several years of attempting to pass it into law.

A second general category of pertinent published material was a body of periodicals published by organized groups who favored a stronger food and drug law, if not always the Copeland bills. Among the more fruitful periodicals, from the view of organized medicine, were the *Journal of the American Medical Association, Journal of Public Health,* and *Hygeia.* A number of publications by the various national women's organizations proved beneficial. The best source from this perspective was the *Journal of Home Economics*. The posture of the professional pharmacy groups who early came to favor a stronger law was best exemplified in the *Journal of the American Pharmaceutical Association*. Also of significant value were the *American Journal of Pharmacy* and the *American Journal of Pharmaceutical Education*. The view of the more militant consumer groups was well presented in the *General Bulletins* of Consumers' Research. Unfortunately the latter were very difficult to locate for the decade of the thirties. For purposes of this study the *Bulletins* were graciously loaned to the author by the Consumers' Research organization from its private files in the New York City office.

A third category of source material was a very large number of lay periodicals. Coverage in such journals was sporadic and abbreviated but in total they supplied both factual data and editorial comment which proved invaluable to analysis of the long struggle for revision. Most useful were the *New Republic, Nation, Christian Century, Business Week, Commonweal, Forum,* and *Time*. Many additional periodicals consulted offered occasional but important coverage. Typical of such journals were *Survey, North American, Scholastic, Literary Digest, Good Housekeeping,* and *Ladies' Home Journal*. A beginning reference point here, of course, was the *Readers' Guide to Periodical Literature*. On several of the most useful journals, however, this writer found that coverage and comment of revision efforts were integrated into small news columns and did not show up in the *Readers' Guide*.

A final group of publications vital to the story was a variety of trade journals which presented the industry point of view. The position of advertising was ably and consistently expressed in *Printers' Ink* and *Advertising and Selling*. The journal *Food Industries* proved to be the best continuing statement of view for the food manufacturers. Three of several good references to the voice of the drug and cosmetic interests were *Drug and Cosmetic Industry, Oil, Paint and Drug Reporter,* and *Standard Remedies*. The latter was particularly valuable in that it directly reflected the position of the powerful Proprietary Association. While the other trade publications were easily located, *Standard Remedies* was difficult to find. This writer found the only complete file in the library of the Department of Agriculture, Washington, D.C.

MISCELLANEOUS MATERIAL

A number of miscellaneous items, all primary in nature but which do not lend themselves to the previous categories, also require mention. Under the auspices of an oral history project sponsored by the National Library of Medicine and the Public Health Service the author was provided the opportunity to do taped interviews with Rexford G. Tugwell, Arthur Kallet, and Dr. Morris Fishbein, retired editor of the *Journal of the American Medical Association*. Transcripts of these interviews are located in the National Library of Medicine, Bethesda, Maryland, and Emory University Library, Atlanta, Georgia. As previously noted in connection with the *Congressional Record*, Charles W. Dunn in his *Federal Food, Drug and Cosmetic Act* (New York, 1938) reprinted all pertinent debate on what would become the 1938 law. Dunn's volume also included the successive versions of the Copeland bill for both the House and Senate along with many other Congressional documents relative to the reform battle. Dunn has further published a second work similar in nature, *The Wheeler-Lea Act: A Statement of Its Legislative Record* (New York, 1938). This volume proved quite valuable since at points the evolution of the Wheeler-Lea statute was tightly woven into the history of the 1938 food and drug law.

Two complete issues of the legal journal *Law and Contemporary Problems* were highly profitable. Under the editorship of David F. Cavers the December 1933 issue (Volume I) and the Winter 1939 issue (Volume VI) were devoted to a symposium on governmental control over the food, drug, and cosmetic market. The former volume provided a variety of views on Senate bill 1944, the latter a well-rounded series of evaluations of the newly passed 1938 statute. Finally, three articles by contemporaries of the reform struggle warrant notice. One was Rexford G. Tugwell's "The Preparation of a President," *Western Political Quarterly*, I (Winter 1948), pp. 131-53, which makes several telling points about Franklin D. Roosevelt's early feeling on a new drug law. The second was Paul B. Dunbar's "Memories of Early Days of Federal Food and Drug Law Enforcement," *Food, Drug, Cosmetic Law Journal*, XIV (February 1959), pp. 87-139, which also provided valuable information on the decision to try for a new law. The third was T. Swann Harding's "The Battle for a Better Food and Drug Law," *American Journal of Pharmacy*, CXVIII (October 1946), pp. 338-59. It contained a pertinent evaluation of the different interest groups involved in the reform fight.

INDEX

Adams, Samuel Hopkins, 18
advertising: inadequate
  regulation over, 5, 6, 7;
  criticized in guinea pig
  volumes, 19; proposed controls
  in S.1944, 28-29; use of
  "red clause," 40, 104; use of
  "pink slips," 104; criticized in
  Our Master's Voice, 110;
  abuses displayed, 126-28;
  controls in Wheeler-Lea Act,
  172-74. *See also* advertising
  controls; advertising industry
advertising controls: need for
  new, 5-6, 7; APMA asks new
  federal, 15; in S.1944, 28-29;
  issue in 1934, 62; Robb on,
  79; industries view issue of,
  82-83, 90-92; Bailey opposes
  FDA, 83; issue in S.5 (35),
  90-94, 95, 116-18; FDR
  equivocal on location of, 94,
  118-19, 122; Campbell asks
  FDA, 103; drug trade supports
  FTC, 105; in House version
  S.5 (35), 120; industries divide
  on House S.5 (35), 121, 122;
  consumers dislike House S.5
  (35), 121, 122; Campbell
  opposes House S.5 (35),
  121-122; AMA objects to House
  S.5 (35), 121, 122

  Senate-House compromise on,
  122-23; issue in S.5 (35), death,
  123-24; and FDA display,
  126; FDA seeks by amendment,
  132; in S.5 (37) and H.R. 300,
  135; trade reaction to S.5
  (35)-H.R. 300 provisions on,
  136; and Lea bill, 146-47, 148;
  pharmacy divided on issue of,
  147-48; FDA control proponents
  silent on, 148; and House
  version S.1077, 149; issue
  renewed in 1938, 171; and

House debate on S.1077,
  171-72; Copeland on issue of,
  173; FTC gains new, 173-74;
  FDA denied, 193; consumers
  and FTC, 206n; FDR and issue
  of, 209; mentioned, 206, 217.
  See also advertising; advertising
  industry
Advertising Federation of
  America, 50, 83. *See also*
  advertising industry
advertising industry: criticized by
  Tugwell, 36; mixed reaction to
  S.1944, 38; and press coverage
  of S.1944, 39-42; and "red
  clause," 40; reacts to S.2800,
  52, 53-54, 70; reaction to
  Huddleston bill, 69; reaction to
  S.5 (35), 80, 82; opposes FTC
  advertising controls, 91-92;
  Chase comments on, 111; House
  version S.5 (35) and comment
  on, 120; reaction to S.5 (37)
  and H.R. 300, 136; reaction to
  1938 food law, 196-97; role in
  food law fight, 214-15. *See also*
  advertising; advertising controls
Agricultural Adjustment Act, 202
Agriculture, U.S. Department of:
  proposed powers criticized, 34;
  report of Sulfanilamide disaster,
  167, 168; and S.5 (37) review
  provisions, 178, 181-82, 186,
  188; and spray residue, 180;
  attacked by Lea, 186;
  mentioned, 15, 131, 135, 142.
  See also Henry Wallace
Aitchison, D., 37
Alice in Wonderland, 22, 31
Allen, Robert, 118
Alpher, Isidore, 19, 113
Alvarez, Walter, 19, 55-56
Ambruster, Howard K., 9, 10, 11,
  12, 86, 166

143, 193. *See also* multiple seizure

Bailey, Josiah: supports Beal bill, 68; opposes FDA advertising controls, 83; prepares S.5 (35) amendment, 95; fears reaction to amendment, 98-99; seeks FDR's amendment reaction, 99; accepts amendment compromise, 100; urged to reject S.5 (37) court review, 187; mentioned, 93, 94, 95. *See also* Bailey amendment

Banbar, 44, 96

Barber, Mrs. Alvin, 103

Barnes, Harry Elmer, 86

Beal bill, 68

Beal, James, 31, 34

Bealle, Morris, 37

Berle, Adolph, 133

Beta Lambda, 110

Bituminous Coal Act of 1937, 192

Black, Loring, 68

Blair, Frank, 30, 67, 89, 92

Blakeley's Acid Iron Material, 127

Boland bill, 69

Brindze, Ruth, 19, 113

Bristol, Lee, 49-50, 164

Brown, Mrs. LaRue, 133-34

Bureau of Chemistry, 11

Burton, Laurence, 35

Business Week: on S.1944, 29; on advertising control issue, 90; on *American Chamber of Horrors*, 112; fears FTC advertising victory, 117; on Tugwell's resignation, 133; on Copeland-FDR differences, 142; on Lea and FDA, 145

Byrns, Joseph, 118, 121, 139

Calhoun, Dr. A. S., 157-58

Campbell, Walter: visits Tugwell, 3; asks investigation of FDA, 13; defends FDA, 14-15; feelings toward Copeland, 16, 60; sees press blackout on S.1944, 43; curtails FDA promotion of S.1944, 47; foresees revision of S.1944, 48;

and S.2800, 57; on Chamber of Horrors and Crazy Crystals, 58; on FDA advertising needs, 103, 116; rumored conflict with Tugwell, 106-107; and increase of FDA promotional work, 107; condemns House version S.5 (35), 120, 122; and FDA display on advertising abuses, 126; on drug reform by amendment, 131-33; and Tugwell's resignation, 133-34; on Sulfanilamide deaths, 156; on Dunn's drug control plan, 169-70; rejects House version S.5 (37). *See also* FDA

Canopus, 75

Carroll, Frank, 166

Carroll, Lewis, 21

Casey's Compound, 127

Cavers, David: and drafting S.1944, 24; defines "Tugwellomania," 36; on court review issue, 180; on 1938 food law, 199; on opposition to new food law, 213; evaluates 1938 food law, 219-20

"Chamber of Horrors": explained, 44; reaction to, 44; use curtailed, 47-48; limited use in 1934, 58; Crazy Crystals withdrawn from, 58; mentioned, 104, 128

Chapman, Virgil: enters food law fight, 102, 103; on S.5 (35) House hearings, 102, 104; visits Chamber of Horrors, 103-104; probes "red clause" and "pink slip," 104; on Institute of Medicine Manufacturers, 104; on Lydia Pinkham's Compound, 104; House strategy, S.5 (35), 105; trade anger at strategy, 106; and House advertising provisions, 120-21; food law amendment approach, 134; introduces H.R. 300, 135; S.5 (37), House committee role, 146; on Lea bill, 148; fears future of S.5 (37), 148-49;

new law promotion criticized, 43-44; new law promotion curtailed, 47-48

Deficiency Appropriations Act, effects of, 57-58; pleased with women's organizations, 66; trades fear anger of, 73-74; reaction to S.5 (35), 76; criticized by Medical Society of New Jersey, 85; on Bailey-Copeland compromise, 100; resumes promotional activities, 107-108; blamed for *American Chamber of Horrors*, 108, 112; advertising control needs defended, 116; seeks FDR's aid, 118; and House version S.5 (35), 120, 121-22; House denies advertising powers, 120; linked to S.5 (35) death, 123-24; display of advertising abuses, 126-29; backed by CR, 130-31; suggests amending 1906 law, 131-34; receives S.5 (37) text, 135; trade support on advertising, 136; critical of first version S.5 (37), 141; and FDR's action on S.5 (37), 142-43; Lea unreceptive to, 145; advertising backing by American Association of Colleges of Pharmacy, 147; advertising proponents silent, 148; minority House committee support, 148; investigates Elixir Sulfanilamide, 154; inspectors reach Massengill plant, 156; 1906 law hinders actions, 156-57, 160-61, Sulfanilamide seizure problems, 157-60; Sulfanilamide deaths reported to, 159; efficiency in Sulfanilamide seizures, 160; case against Massengill, 160-61; Sulfanilamide and support for, 162-64; Sulfanilamide publicity valuable to, 165-66; criticized by Ambruster, 166;

drafts new drugs amendment, 168; favors H.R. 9341, 168-69; trades seek concessions from, 170; seeks veto on S.1077, 172-73; fears passage S.1077, 173; defeated on advertising issue, 174; S.5 (37) review provisions affects, 178, 179-80; and spray residue issue, 179; condemns S.5 (37) review provisions, 181; drafts court review compromise, 184-85; final version S.5 (37) pleases, 191; operation compared to FTC, 191-92; 1938 law denies advertising controls, 193; and multiple seizure in 1938 law, 193; initial problems of 1938 law, 194; Tugwell's interest in, 203; changes in nature of, 204-205; Young on, 207; originated new law movement, 207; women most effective support, 213; administration of new law, 220

Food, Drug, and Cosmetic Act of 1938: court review provisions analyzed, 191-93; weaknesses in, 193-94; provisions compared to 1906 law, 195-96; trade reaction to, 196-97; CU and Tugwell critical of, 197; professional pharmacy pleased with, 198; women applaud, 198; Wallace satisfied with, 199; press coverage of fight for, 199; ends New Deal, 201, 219; and First New Deal, 201-205; and business community, 204, 220; and Second New Deal, 205-206; consensus legislation, 206; competition in fight for, 206-207; origin outside New Deal, 207; and New Deal priorities, 207-208; political factors in drive for, 208; and FDR's role, 208-209; effects of FDR-Copeland differences,

Campbell comments on, 103; S.5 (35) opposition to, 113; provision in House S.5 (35), 119; Campbell seeks to retain, 132; S.5–H.R. 300 provisions on, 135; women seek unrestricted, 138; in Senate S.5 (37), 141; S.5 (37) provisions modified, 143; in House-Senate Conference S.5 (37), 190; FDA and 1938 food law, 193; mentioned, 206, 217. *See also* Bailey amendment

Musser, Mrs. J. W., 109

Nation, 42, 111, 213
National Advisory Council of Consumers and Producers, 39, 39n, 78
National Association for the Advancement of Colored People, 208
National Association of Boards of Pharmacy, 72, 129
National Association of Insecticide and Disinfectant Manufacturers, 164
National Association of Retail Druggists, 36, 39, 70
National Better Business Bureau, 11
National Broadcasting Company, 78
National Congress of Parent-Teachers Association, 46. *See also* women's organizations
National Consumers League, 187
National Drug Trade Conference: defined, 31n; reacts to S.1944, 31, 39; backs McCarran-Jenckes bill, 68; support for S.5 (35), 129; mentioned, 71, 72. *See also* drug industry
National Editorial Association, 34. *See also* advertising industry
National Industrial Recovery Act, 202, 203

National League of Women Voters, 110, 133, 187. *See also* women's organizations
National Liberties Association, 37, 37n
National Publishers Association, 39, 83. *See also* advertising industry
National Pure Food and Drug Congress, 4
National Recovery Administration, 33, 50, 51, 203
National Wholesale Druggists Association, 70, 136, 215. *See also* drug industry
national women's organizations, *see* women's organizations
Net Weight Act of 1913, 8
New Deal: effects of S.1944 on, 39; Copeland and, 63, 76, 119, 133, 139-40, 142, 144, 147; court revision plan, 140; and S.5 (37) Senate strategy, 140-41; FTC and, 144-45; and Wheeler, 173; duration of, 201, 219; Tugwell and First, 201-203; 1938 food law and First, 201-205; 1938 law typical legislation, 204; 1938 food law and Second, 205-206; affects food law struggle, 207; priorities and new food law, 207-208; Tugwell on priorities of, 207-208; politics and new food law, 208; consumer interests in, 210-11; and consumer-oriented society, 212; rejects Tugwell's philosophy, 219; Paul Conkin on, 219; and business community, 220
New Despotism, The, 33-34
New Orleans Picayune, 162
New Republic: on opposition to S.2800, 52; on S.2800, 56; on advertising control issue, 90; on guinea pig muckraking, 111; on Sulfanilamide disaster, 162; and 1938 food law, 213
Newsweek, 165

[248]